Asian Crops and Human Dietetics

HAWORTH FOOD & AGRICULTURAL PRODUCTS PRESS™
Crop Science

The Lowland Maya Area: Three Millennia at the Human-Wildland Interface edited by A. Gómez-Pompa, M. F. Allen, S. Fedick, and J. J. Jiménez-Osornio

Biodiversity and Pest Management in Agroecosystems, Second Edition by Miguel A. Altieri and Clara I. Nicholls

Plant-Derived Antimycotics: Current Trends and Future Prospects edited by Mahendra Rai and Donatella Mares

Concise Encyclopedia of Temperate Tree Fruit edited by Tara Auxt Baugher and Suman Singha

Landscape Agroecology by Paul A. Wojtkowski

Concise Encyclopedia of Plant Pathology by P. Vidhyasekaran

Molecular Genetics and Breeding of Forest Trees edited by Sandeep Kumar and Matthias Fladung

Testing of Genetically Modified Organisms in Foods edited by Farid E. Ahmed

Fungal Disease Resistance in Plants: Biochemistry, Molecular Biology, and Genetic Engineering edited by Zamir K. Punja

Plant Functional Genomics edited by Dario Leister

Immunology in Plant Health and Its Impact on Food Safety by P. Narayanasamy

Abiotic Stresses: Plant Resistance Through Breeding and Molecular Approaches edited by M. Ashraf and P. J. C. Harris

Teaching in the Sciences: Learner-Centered Approaches edited by Catherine McLoughlin and Acram Taji

Handbook of Industrial Crops edited by V. L. Chopra and K. V. Peter

Durum Wheat Breeding: Current Approaches and Future Strategies edited by Conxita Royo, Miloudi M. Nachit, Natale Di Fonzo, José Luis Araus, Wolfgang H. Pfeiffer, and Gustavo A. Slafer

Handbook of Statistics for Teaching and Research in Plant and Crop Science by Usha Rani Palaniswamy and Kodiveri Muniyappa Palaniswamy

Handbook of Microbial Fertilizers edited by M. K. Rai

Eating and Healing: Traditional Food As Medicine edited by Andrea Pieroni and Lisa Leimar Price

Physiology of Crop Production by N. K. Fageria, V. C. Baligar, and R. B. Clark

Plant Conservation Genetics edited by Robert J. Henry

Introduction to Fruit Crops by Mark Rieger

Generations Gardening Together: Sourcebook for Intergenerational Therapeutic Horticulture by Jean M. Larson and Mary Hockenberry Meyer

Agriculture Sustainability: Principles, Processes, and Prospects by Saroja Raman

Introduction to Agroecology: Principles and Practice by Paul A. Wojtkowski

Handbook of Molecular Technologies in Crop Disease Management by P. Vidhyasekaran

Handbook of Precision Agriculture: Principles and Applications edited by Ancha Srinivasan

Dictionary of Plant Tissue Culture by Alan C. Cassells and Peter B. Gahan

Handbook of Potato Production, Improvement, and Postharvest Management edited by Jai Gopal and S. M. Paul Khurana

Carbon Sequestration in Soils of Latin America edited by Rattan Lal, Carlos C. Cerri, Martial Bernoux, Jorge Etchevers, and Eduardo Cerri

Genetically Engineered Crops: Interim Policies, Uncertain Legislation edited by Iain E. P. Taylor

Management of Nematode and Insect-Borne Plant Diseases edited by Geeta Saxena and K. G. Mukerji

Asian Crops and Human Dietetics

Usha R. Palaniswamy

The Haworth Press
Taylor & Francis Group
New York and London

For more information on this book or to order, visit
http://www.haworthpress.com/store/product.asp?sku=5504

or call 1-800-HAWORTH (800-429-6784) in the United States and Canada
or (607) 722-5857 outside the United States and Canada
or contact orders@HaworthPress.com

Published by

The Haworth Press, Taylor & Francis Group, 270 Madison Avenue, New York, NY 10016.

Cover design by Kerry Mack.

Library of Congress Cataloging-in-Publication Data

Palaniswamy, Usha R.
Asian crops and human dietetics / Usha R. Palaniswamy.
p. cm.
ISBN: 978-1-56022-312-2 (hard : alk. paper)
1. Food crops—Health aspects. 2. Medicinal plants. 3. Cookery, Asian. I. Title.

SB175.P355 2008
613.2095—dc22

2007021433

Invocation

Om kreem parasakthiye namaha
Thriyambagam yeja mage suganthim
Pushti varthanam
Urvarugamiva bandhanath mrithyo mukshi
Yama mridhathu

I worship the three-eyed Lord Shiva who is full of sweet fragrance and nourishment for human beings. May He liberate me from bondage and death as the cucumber is severed from the vine and lead me to immortality.

Dedicated to my parents who gave me life
Indirani and K. M. Palaniswamy
And to my daughters to whom I give my life
Archana and Madhushree

ABOUT THE AUTHOR

Usha Palaniswamy, PhD, MEd, is the current Chair, Division of Natural Sciences and Mathematics, Excelsior College, Albany, NY. Her current research focuses on international farming systems, integrated and sustainable farming practices, Asian traditional medical systems, Asian medicinal plants, human issues in horticulture, critical health issues of Asian Americans, and examinations of phytochemicals that promote human health and prevent diseases. She is the author of over 40 peer-reviewed research and invited review journal articles, several book chapters, and the books *A Guide to Medicinal Plants of Asian Origin and Culture* (CPL Press, UK) and *Handbook of Statistics for Teaching and Research in Plant and Crop Sciences* (Haworth). She is also Lead Editor of *Acta Horticulturae: Traditional Medicine and Nutraceuticals;* and Editor of the *Journal of Herbs, Spices and Medicinal Plants*. Dr. Palaniswamy received several awards for her teaching, research and service efforts: President's Volunteer Service Award from the US President's office in 2006; University of Connecticut's General Education grant for 2004-2005; Outstanding Young Scientist Award from the Association of Agricultural Scientists of Indian Origin in 2004; Excellence Award for Teaching Innovation from the American Association of University Professors and an official citation from the State of Connecticut general assembly for her innovations in teaching in 2002. She is also an international research and teaching fellow of the Visvabharathi University, West Bengal, India; Bharathiyar University, Coimbatore, India; the International Rice Research Institute, Philippines; and the Prem Nath Agricultural Science Foundation, Bangalore, India. She is currently the North East District delegate and a Board member of the Herb Society of America. She was faculty of the Asian American Studies Institute, University of Connecticut 1999-2007.

CONTENTS

Acknowledgments

I am extremely thankful for the support I received from my friends and family, especially my daughters, who worked with me in the preparation of this manuscript.

doi:10.1300/5504_a

Introduction

All humans need a diet that provides sufficient energy and nutrients for performing metabolic functions necessary for living. Although needs and requirements are similar for all of us, and we all achieve the same end in this process, peoples, cultures, families, and individuals fulfill this basic need in diverse ways. Epidemiological studies documenting variations in public health in different parts of the world almost always point to the fact that health and disease are related to not only an individual's genetics but also environmental conditions, including regional climatic conditions, food and agricultural practices, social and economic conditions, preventive health care and immunization programs, public hygiene, educational status, and availability of uncontaminated water as well as the interaction of these factors and the individual's personal choices and behaviors. Detailed examination of these practices often yields interesting analyses of human behavior in seeking food for the promotion of health and wellness of mind and body.

Migration of populations out of their native lands has been caused by poverty, religion, societal conflicts, and politics. Asian peoples' migration started several hundred years ago, and they have spread to all countries of the world. For example, the earliest migration of people from the Indian Subcontinent was to Southeast Asia, where small groups settled and merged with the larger local society. Later, during the nineteenth century and until the end of the British rule, migration occurred in various forms for various reasons: to Mauritius, Guyana, the West Indies, Fiji, and East Africa as slave labor under the indenture system; to Sri Lanka, Burma, and the Malay regions (Malaysia and Singapore) to work on tea and rubber plantations; to the United Kingdom, Australia, and New Zealand after independence in 1947 for personal reasons related to family and education; and to Persian Gulf countries after the 1970s oil boom in the Middle East for lucra-

doi:10.1300/5504_b

tive jobs. The United States became a favored destination after the 1965 change in U.S. immigration law. The United States was initially considered the "land of the whites" and not considered appropriate for raising children or maintaining traditional Indian culture. But the United States became the land of opportunity in the mid-1980s as evidenced by the increased migration from India to the United States. China has been and continues to be a major source of international migration. It is speculated that people of Chinese ancestry live in virtually every country of the world today; it has been estimated that 33 million ethnic Chinese live outside China, Taiwan, and Hong Kong. Thus, Asian cultures and cuisines are no longer limited to a single group or region, and Asian foods, ingredients, and preparation styles are becoming part of American mainstream culture.

In North America, food and medicine needs are more diverse and becoming more complex than in any other part of the world because of shifting demographics caused by migration. Further change will come as minority populations and ethnic diversity grow and are combined with scientific findings supporting the benefits of a traditional diet. Asian Americans are one of the fastest-growing minority populations in the United States. The Asian population is the fastest-growing group in all regions of the United States. Asians are the fourth largest of the race in all regions of the United States except the Western region where they rank third. In the United States, the Asian-American population is projected to have the greatest gains in the West with an increase of 7 million persons and in the Northeast with an increase of 2 million (Campbell 1996). The Asian-American population also comprises a growing diversity, deriving from more parts of Asia than ever before. Projections from the census of 2000 and immigration trends show that by 2050, if not sooner, no single group will constitute more than 50% of the total population in the United States. This presents mixed challenges in all dimensions of life but particularly in the health care issues and policy formation that are vital to a healthy community. Health care must adapt to encompass findings from numerous epidemiological, biomedical, and clinical studies documenting the benefits of ethnic and so-called traditional diets, accommodate a growing need for global perspectives, and meet the increasing complexity of approaches in achieving basic needs. These approaches now include the areas of agricultural, food, and medical sciences as well as biomedical technologies.

The modern emphasis on prevention rather than treatment and the attitude that "life is long, live it well" promote a healthy lifestyle and bring Asian foods to the forefront. Americans are reducing fat in their diet and being selective in the kinds of carbohydrates, fats, and proteins they eat. They are recognizing the importance of vegetables and fruits in health promotion and striving to achieve the recommended five-a-day portions. In this process they look for taste, appearance, and variety, and Asian foods are popular choices because their distinct spices and flavorings create a unique blend offering intense flavors and varied textures, visual appeal, and a variety of nutritious ingredients found in no other cuisine. A prevailing perception based on strong scientific evidence is that consumption of Asian spices, lentils, and soy products, which contain a number of phytochemicals, is a natural method for curing ailments, managing chronic conditions, promoting health, and preventing disease. The growing popularity of and demand for alternative medicine in mainstream America has brought an increased awareness of ancient Asian medical systems (traditional Chinese medicine and the Indian Ayurvedic, Unani, and Siddha forms of medicine), which are perceived as natural ways of achieving wellness of body and mind. These ancient systems of health care are holistic systems of medicine that use a constitutional model and aim to provide guidance regarding food and lifestyle so that healthy people can continue to stay healthy and people with health challenges can improve their health. These medical systems' concept of food as medicine has triggered interest in functional foods and the formulation of nutraceuticals. The principles of moderation and variation in diets, individualization of dietary recommendations, and the inclusion of physical activities in daily life advocated by these systems as ways of life are scientifically sound and clearly reflected in modern-day Western dietary guidelines.

A brief historical perspective on human diets and dietetics will illuminate the past; give meaningful depth, breadth, and perspective to the present; and provide valuable direction for future progress in the study of dietetics and related fields. Historically, early humans settled closer to water sources, resorted to hunting and gathering, and adopted different methods of processing and preserving so that fresh-water food and seafood almost always were part of a diversified diet. Human diets included naturally occurring vegetation, wild animals, and aquatic resources, and vigorous physical exertion was required to acquire food. However, rapid urbanization and global migration of much of today's world population has distanced people from not only their countries of origin but also the origin of their food, as well as from an understanding and acknowledgment of the diversity required in one's diet. In the modern world, access to a diversified food supply for a given population depends on regional and local biodiversity and whether trade routes for desired supplies have been established. Health benefits of a diversified diet have been documented; At all stages of life human physiology requires a wide spectrum of food components, macronutrients, micronutrients, and phytochemicals. A diet's components must be chosen carefully to provide all needed nutrients in adequate amounts, proper proportions, and appropriate cultural context. The amount of each nutrient required depends on an individual's age, body weight, physiology, and metabolism. Other factors to consider include the quality of dietary ingredients, effects of cooking and processing, and the bioavailability of nutrients in specific combinations.

Although as early as 1785 several scholars had ideas about how food was used in the body, it was not until the next century that the main elements of food were identified by chemical analyses in France and scientists started to test their ideas using quantitative scientific methods. In 1817 Francois Magendie demonstrated with experiments on dogs that food should be able to provide all the needs of a living being. Magendie's guidelines on "diversity and multiplicity" and use of animal species as models to test nutrients and foods for their effects on humans continue to resonate in modern-day dietary guidelines and research methods (Magendie 1831). Speculation on the importance of plant foods in human nutrition came as early as 1806 (Vauquelin and Fourcroy 1806), and "synthesis only by plants" was hypothesized by Dumas (1841). The importance of fruits and vegetables in the human diet was again demonstrated when several researchers documented curing scurvy by a diet of fresh vegetables and fruit juice (Baly 1843; Bryson 1850).

Although the ancient medical and healing systems of humans such as the traditional Chinese and the Indian Ayurvedic systems have taken a dietary approach to human health and wellness for more than 5,000 years, a science-based approach to achieving a healthful diet began in the United States about a hundred years ago when Wilbur Olin Atwater, as the first director of the Office of Experiment Stations in the U.S. Department of Agriculture (USDA), initiated educating the public about food choices and developing a food guide. Atwater, known as the father of nutritional science in the United States, encouraged basic nutritional research in agricultural experimental stations. Atwater focused on protein analysis and proposed that meat protein could be safely replaced by that from beans. In 1894 he published tables of food composition and dietary standards for the U.S. population (Atwater 1894). The first food tables provided data on the protein, fat, carbohydrate, ash (mineral matter), and "fuel" value for some commonly available foods, to guide people in choosing food with the necessary calories.

Atwater recognized the need for a well-balanced diet, and he connected food composition and dietary intake with health. The first food guide that could translate recommendations on nutrient intake into recommendations on food intake was first developed by Caroline Hunt (1916), who divided food into five groups—milk and meat, cereals, vegetables and fruits, fats and fatty foods, and sugars and sugary foods—on the basis of what was known then about nutritional needs and food composition. In 1921 a guide for the average family was released that used these same food groups, and it suggested the amounts of food to purchase each week. In the early 1930s Hazel Stiebeling, a food economist in USDA's Bureau of Home Economics, developed buying guides for food shoppers that defined the amounts of foods to buy each week, at four cost levels, that would meet the nutritional needs of men, women, and children of different ages. Stiebeling outlined 12 major food groups: (1) milk; (2) potatoes and sweet potatoes; (3) dry beans, peas, and nuts; (4) tomatoes and citrus fruits; (5) leafy green and yellow vegetables; (6) other vegetables and fruits; (7) eggs; (8) lean meat, poultry, and fish; (9) flours and cereals; (10) butter; (11) other fats; and (12) sugars. Later in 1946, the USDA announced the Basic Seven nutritional guidelines, which were a revision of the wartime nutritional guidelines (USDA 1946): (1) green and yellow vegetables; (2) oranges, tomatoes, and grapefruit; (3) potatoes and other vegetables and fruit; (4) milk and milk products; (5) meat, poultry, fish, eggs, and dried peas and beans; (6) bread, flour, and cereals; and (7) butter and fortified margarine. Because of the complexity of the Basic Seven recommendations, it was later condensed to the Basic Four: milk, meats, fruits and vegetables, and grain products (USDA 1958).

In the late 1900s the USDA addressed the prevalence of unhealthy foods and the rise in chronic health conditions, degenerative diseases, cancer, stroke, and heart disease. In the food guide pyramid released in 1992 (USDA 1992), both the graphic and the text conveyed the concepts of variety and proportionality. A new food pyramid graphic released in 2005 conveys four major ideas: variety, proportionality and moderation, physical activity, and ways to individualize the recommendations (Myfoodpyramid 2005).

Similar pictorial food guidelines, as shown in Figure I.1, have been formulated internationally because people learn far more from looking at simple pictorial representations than they do from reading pages of information. The pyramid shape used by the United States is not necessarily the graphic of choice for food guides worldwide because of cultural issues and differences in ways of communicating. For example, Australians have adopted a circular plate, the Japanese a spinning top, Indians a series of ascending steps, and the Thai an inverted pyramid. The different graphics depict cultural norms and symbols, an emphasis of dietary guidelines specific to each country, and the prevailing health status and the social and economic conditions of its population. In developed countries where deficiency is not an issue, food guides tend to promote a diet that prevents chronic disease. However, in developing countries, where poverty and poor health can coexist, the message needs to address both deficiency and chronic health conditions. Despite the different shapes and colors, all the pictorial representations communicate similar messages: energy balance, diversity in food choices, and moderation in quantity. The largest segment of the pictorial is sources of carbohydrates—rice, noodles, breads and cereals—vegetables and fruits are the next largest group, followed by meat and dairy products, and then foods to be consumed occasionally or minimally. Use of color to represent the categories is also common. For example, Canada's Food Guide to Healthy Eating is a four-banded rainbow, with each color representing one of its four food groups and the width of the rainbow bands varying with the dif-

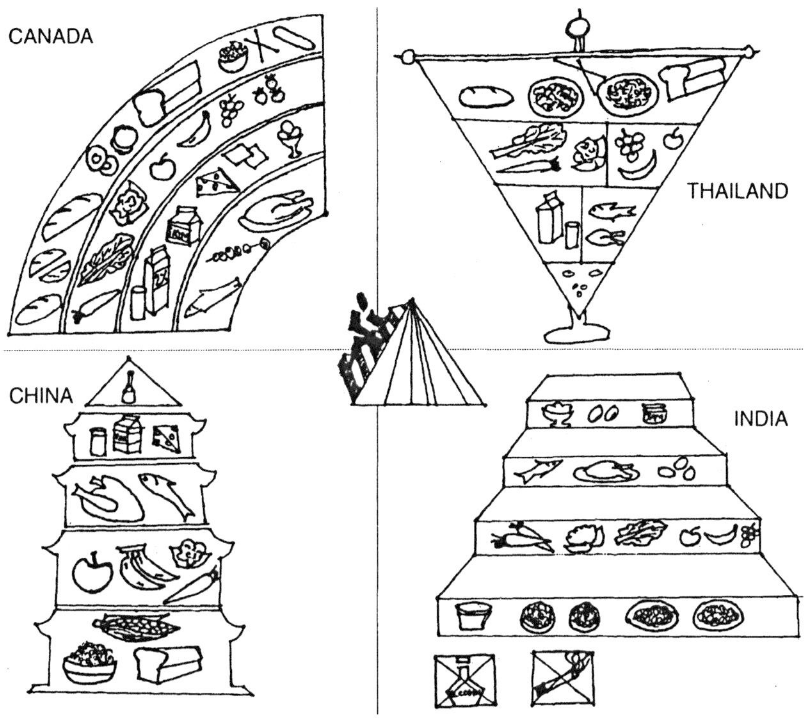

FIGURE I.1. Selected International Pictorial Dietary Guidelines Formulated for Education of the Public

ferent amounts to be consumed from each group. The United States also adopted the use of color in its most recent food pyramid, with a color representing each food group and the size of the pyramid slice representing the proportion of the food group in the diet. In China, a pagoda is the food guide, with the cereal group forming the base of the pagoda and narrowing levels for the rest of the food groups (Painter et al., 2002).

From a dietitian's point of view, good nutrition means accounting for the required calories and other essential nutrients in the daily diet. The human body is composed of 54% to 62% water, 15% to 17% protein, 14% to 25% fat, 5% to 6% mineral matter, and about 1% carbohydrate. The human body also requires 17 vitamins, 24 mineral elements, and water. The energy yield per gram of macronutrients is carbohydrate, 4 kcal; protein, 4 kcal; and fat, 9 kcal (by convention, in nutritional measurements 1 food calorie = 1 kcal).

Although carbohydrates represent the smallest proportion of the human body, they make up the bulk of our diet since they constitute the chief source of energy. Animal cells use oxygen to burn carbohydrates to produce energy; carbohydrates also help utilize proteins and fats for more complex physiological functions than production of energy. Carbohydrates consumed in excess of the body's needs are converted to fat and glycogen and stored for use when needed. Starches and sugars are the main sources of energy. There are two types of polysaccharides from plant foods: the storage starch polysaccharide and the nonstarch polysaccharide (NSP), or dietary fiber. Cereal grains; pulses and other seeds; and root and stem vegetables such as potato, sweet potato, cassava, and plantains contain large amounts of starch, constituting the world's most important starchy foods. In addition, sugarcane, beets, and fruits and vegetables that are sweet provide sugar. The proportions of starch, sugar, and dietary fiber in the diet are highly variable for people of different communities in the world. Since the role of dietary fiber in human health has been scientifically documented and has gained general acceptance among the public (European Cancer Prevention Consensus Panel 1998; Hill 1997; Murtaugh et al. 2003; Pereira et al. 2003; Terry et al. 2001), the American Dietetic Association (ADA) recommends 25 g dietary fiber per day for a 2,000 kcal/day diet or 30 g/day for a 2,500 kcal/day diet (American Dietetic Association 1997).

Proteins are building blocks for the body; they are important as structural constituents of cellular membranes, essential in maintaining and repairing worn tissues, and major constituents of enzymes necessary for various biochemical reactions. Plant sources of protein include pulses and other grains such as oil seeds and cereals. In general, nuts and seeds are rich sources of storage proteins. Although proteins are not the primary source of energy, the body can convert protein into energy in the absence of adequate carbohydrates and fats in the daily diet.

Lipids are the most concentrated source of food energy, furnishing more than two times the energy of either carbohydrates or proteins. They contain less oxygen and more hydrogen than carbohydrates. In plant foods fats are mainly confined to membranes or may be present as stored fat globules within the cell. Fats are concentrated in oil seeds such as peanut, sunflower, and safflower; in fruits such as avocado and olive; and in the germ of cereals.

Vitamins are essential nutrients required for the proper utilization of nutrients and are responsible for the maintenance of good health. Vitamins are the constituents of enzymes and function as catalysts in the regulation of most biochemical processes. A number of vitamins (A, C, and E) and their precursors are antioxidants, which protect the body from damage by charged oxygen molecules and help maintain the immune system. Plants are excellent sources of vitamins, whose concentrations vary with the conditions of growth; stage of maturity; growing conditions; and handling, processing, storage, and cooking methods.

Minerals act as catalysts in many biochemical reactions of the body. Some function in the growth and development of bones (Ca and P) and other structural components, muscular contraction, transmission of messages through nerves, and in the digestion and utilization of other food nutrients. Some minerals—calcium, phosphorus, iron, manganese, and sulfur—are required in large quantities, while others—zinc, copper, iodine, cobalt, and fluorine—are required in smaller amounts. Minerals are widely distributed in plant foods, and their amounts vary according to the medium of growth and a number of other cultural and genetic factors.

Water is second only to oxygen in importance to the body. It is an ideal medium for transportation of dissolved nutrients and elimination of waste products from the body. In addition to water consumed as such and in beverages, water needs are met through water provided by fruits and vegetables and by metabolic water.

At the end of the twentieth century, the importance of a number of anutrient phytochemicals in human health promotion and disease prevention were demonstrated by various researchers. Studies continue to show that disease and dietary habits are closely linked and that several functional foods can prevent or cure diseases and promote positive health. In today's world of escalating health costs and scarcity of resources, it is imperative to produce, preserve, process, and promote consumption of vegetables and fruits which directly affects the inception, progression, and perhaps the ultimate manifestations of almost all chronic disorders (Cannon 1992).

The profession of dietetics has evolved to become a component of allied health sciences, part of basic research and in clinical settings, necessary to food-service management, and a contributor in profound ways to society in general. During the past century the field of human nutrition and dietetics has grown to be recognized as a profession with all accompanying accessories—standards of education and practice, codes of ethics, and professional associations focusing on specialty areas.

Plant sciences have grown in parallel in their examinations of plants for food, nutrition, and health promotion. The past century has seen advances in nutritional and agricultural sciences in both research and practice. The next century may see equally pivotal changes as professionals in these fields respond to human nutritional needs in a changing world with innovative practices and more interdisciplinary approaches in research and practice.

As the current consumption of fruits and vegetables increases to meet the recommended portions, there is a greater need to focus on identifying indigenous and traditional vegetables and fruits in specific diets in different cultures that can be added to the current choices in fresh produce. Researchers will continue to expand their knowledge of plants' intricate workings at the cellular level as well as a whole, investigate plants' physiological and defense responses to environmental changes and the phytochemicals that result, and study plants' role in preventing human chronic illness. Dietitians, biomedical scientists, and laboratory and clinical practitioners will continue to translate their research into practical applications and share outcomes for better human health and quality of life. Modern medical and biomedical research on aging has revealed that changes that were once labeled as an inevitable process and part of growing old, such as high blood pressure, obesity, memory loss, and intellectual impairment, are pathological processes that can be controlled by behavior and nutrition. Thus, there is accumulating evidence from multiple sources (Bernstein et al. 2007; Clements and Ashurst 2006; Rudkowska and Jones 2007; Van Dyk and Sano 2007) that good nutrition can help prevent disease and promote health. This is closely related to plant secondary metabolites. These compounds, termed *nutraceutical* or *phytochemical* and classified as nonessential micronutrients that contribute to human homeostasis and general well-being, play a significant role in the maintenance of health. Scientific examination of food plants to determine their phytochemical composition and identify the organic compounds responsible for these activities has increased (Pisha and Pezzuto 1994). Thus, nutritional science, usually interested in quality and safety of foods, has now moved closer to the issue of health promotion, with an increased attention to diversification of dietary sources to include the secondary metabolites from edible plants. And in 1989 dietitians changed their recommendations for a daily intake to reference intakes. Collectively called the dietary reference intakes (DRIs), the new recommendations are in reference values and include the recommended dietary allowance (RDA), the adequate intake, the tolerable upper intake level, and the estimated average requirement. The primary goal of the new dietary reference values was to not only prevent nutrient deficiencies but also reduce the risk of chronic diseases such as osteoporosis, cancer, and cardiovascular disease.

Vavilov (1940) proposed that the center of origin of a plant species is also the center where one finds the maximum diversity of the species. The eight Vavilovian centers of origin for crop plants and their number of plants are as follows: (1) Chinese, 113; (2) Indian, 117; Indo-Malayan, 55; (3) Central Asiatic, 43; (4) Near-Eastern, 83; (5) Mediterranean, 84; (6) Abyssinian, 38; (7) Southern Mexican and Central American, 45; and (8) South American, 62. The eight primary centers listed comprise only 2% to 3% of the earth's land area. Vavilov speculated that five out of six cultivated crops originated in the old world, and he listed about 640 species of which more than 50% originated in the three centers of origin in the Asian region. Thus, Asian crops and culture have made a significant contribution to human existence. However, some crops cultivated and consumed as an important dietary component in one part of Asia are not known in other parts of the world. The purpose of this book is to close this gap and to bring attention to the crops of Asian origin and culture that have made significant contributions to human dietetics. Most crops highlighted in the chapters are those found in Asian cuisine that have been used historically for health promotion and are documented as being in traditional systems of health and healing. Plant species in this book are those crops currently used only in specific parts of the world and hence underused by humans in general. Such crops have the potential to contribute to diversity in food groups and a greater potential for use as phytonutrients. The foods are grouped in the new food guide pyramid classification that represents the plant kingdom as (Chapter 1) grains, (Chapter 2) vegetables, (Chapter 3) fruits, (Chapter 4) fats and oils, and (Chapter 5) beans and nuts. Chapter 6 discusses important spices that are not generally classified as either vegetable or fruit but are increasingly studied for their chemopreventive properties. Before long these spices might be listed as a separate category in food pyramids and other visual dietary recommendations (Palaniswamy 2001). Chapter 7 deals with herbal preparations, beverages, and substances that different cultures consume after a main meal (e.g., an herbal tea or green tea in China or chewing *pan,* a combination of betel leaves, areca nut, calcium carbonate, and spices and flavorings, in the Indian Subcontinent). The plant species used in the preparation of such digestives and postprandial treats, termed in this book as "subsequents," also are being studied for their potential benefits or consequences in the daily diet.

The challenge ahead is to encourage better lifestyle choices taking into consideration the current epidemiological, medical, and scientific knowledge and making available appropriate plant-based foods in the existing socioeconomic and cultural context in different parts of the world. In achieving this, one needs to develop ongoing evaluation and conservation practices and policies for preservation and utilization of the existing biodiversity of edible crops as well as sound natural resource management practices for development of organic and sustainable farm production systems.

REFERENCES

American Dietetic Association 1997. Health implications of dietary fiber—Position of the ADA. Journal of the American Dietetic Association 97: 1157-1159.

Atwater WO. 1894. Foods: Nutritive Value and Cost. Farmers' Bulletin No. 23. U.S. Department of Agriculture, Washington, DC.

Baly W. 1843. On the prevention of scurvy in prisoners, pauper lunatic asylums etc. London Medical Gazette [new ser.] 1: 699-703.

Bernstein AM, Treyzon L, Li Z. 2007. Are high-protein, vegetable-based diets safe for kidney function? A review of the literature. Journal of the American Dietetic Association 107(4): 644-650.

Bryson A. 1850. On the respective values of limejuice, citric acid, and the nitrate of potash, in the treatment of scurvy. Medical Times and Gazette (London) 21: 212-214, 435-436.

Campbell PR. 1996. Population Projections for States by Age, Sex, Race, and Hispanic Origin: 1995 to 2025. U.S. Bureau of the Census, Population Projections Branch, Population Division, PPL-47. U.S. Census Bureau, Washington, DC.

Cannon G. 1992. Dietary guidelines—Food and health: The experts agree. Consumers' Association Ltd., London.

Clements L, Ashurst I. 2006. Dietary strategies to halt the progression of chronic kidney disease. EDTNA ERCA Journal 32(4): 192-197.

Dumas JB. 1841. On the chemical statics of organized beings. Philosophical Magazine 19: 337-347.

European Cancer Prevention Consensus Panel. 1998. Consensus statement on cereals, fibre and colorectal and breast cancers. European Journal of Cancer Prevention 7(Suppl. 2): s1-s4.

Hill MJ. 1997. Cereals, cereal fibre and colorectal cancer risk: A review of the epidemiological literature. European Journal of Cancer Prevention 6: 219-225.

Hunt CL. 1916. Food for Young Children. Farmers' Bulletin No. 717. U.S. Department of Agriculture, Washington, DC.

Magendie F. 1831. An elementary compendium of physiology. Milligan E, translator. John Carfrae, Edinburgh, UK.

Murtaugh MA, Jacobs DR, Jr., Jacob B, Steffen LM, Marquart L. 2003. Epidemiological support for the protection of whole grains against diabetes. Proceedings of the Nutrition Society 62:143-149.

Myfoodpyramid. 2005. http://www.mypyramid.gov/. Accessed July 18, 2006.

Painter J, Rah JH, Lee YK. 2002. Comparison of international food guide pictorial representations. Journal of the American Dietetic Association 102(4): 483-489.

Palaniswamy U. 2001. Human dietetics and Asian food crops. HortTechnology 11(4): 504-509.

Pereira MA, O'Reilly E, Augustsson K, Fraser GE, Goldbourt U. 2004. Dietary fiber and risk of coronary heart disease: A pooled analysis of cohort studies. Archives of Internal Medicine 164: 370-376.

Pisha E, Pezzuto JM. 1994. Fruits and vegetables containing compounds that demonstrate pharmacological activity in humans. In: Wagner H, Hikino H, Farnsworth NR, editors, Economic and medicinal plant research, Vol. 6. Academic Press, New York. p. 189-226.

Rudkowska I, Jones PJ. 2007. Functional foods for the prevention and treatment of cardiovascular diseases: Cholesterol and beyond. Expert Review of Cardiovascular Therapy 5(3): 477-490.

Terry P, Giovannucci E, Michels KB, Bergkvist L, Hansen H, et al. 2001. Fruit, vegetables, dietary fiber, and risk of colorectal cancer. Journal of the National Cancer Institute 93(7): 525-533.

USDA. 1946. Bureau of Human Nutrition and Home Economics. National food guide. AIS-53, Rev. of NFC-4. U.S. Department of Agriculture, Washington, DC.

USDA. 1958. Food for fitness: A daily food guide. Leaflet No. 424. U.S. Department of Agriculture, Washington, DC.

USDA. 1992. The food guide pyramid. Human Nutrition Information Service. Home and Garden Bulletin No. 252. U.S. Department of Agriculture, Washington, DC.

Van Dyk K, Sano M. 2007. The impact of nutrition on cognition in the elderly. Neurochemical Research 32(4-5): 893-904.

Vauquelin LN, Fourcroy AN. 1806. Memoir upon the germination and fermentation of grains and farinaceous substances. Philosophical Magazine 25: 176-182.

Vavilov NI. 1940. Origin and Geography of Cultivated Plants. (trans. Doris Love).

Chapter 1

Grains

Plants from the grass family (Poaceae or Graminae) are important sources of carbohydrates for the human diet and should make up the largest proportion of a daily diet. In photosynthesis, plants use energy from sunlight to synthesize carbohydrates from carbon dioxide and water, producing oxygen. Plant foods contain easily soluble sugars and two main types of polysaccharides: (1) starch, whose storage form is a glucose polymer bonded by alpha-linkages, and (2) nonstarch polysaccharides, which do not contain alpha-linkages and which make up dietary fiber.

Carbohydrates from plants provide a substantial proportion of the energy in most human diets, which vary by the person's region of residence and economic status. In diets of developing countries, especially in the tropics, as much as 85% energy comes from carbohydrates; in diets of developed countries as little as 30%. Carbohydrates have been traditionally regarded as a simple energy source, but recently their role in human health promotion and disease prevention has been studied in greater detail and they are now recognized as an important functional food that provides additional benefits beyond its nutritional value.

Scientific studies have found that the physiological effects of carbohydrates highly depend on the rate and extent of digestion and absorption in the small intestine and fermentation in the large intestine. Since starch and sugars are mostly broken down by digestive enzymes into monosaccharides and then absorbed in the small intestine, this can lead to elevated levels of plasma glucose and insulin, which are detrimental to health in many contexts, including diabetes, coronary heart disease, cancer, and aging. However, a major portion of dietary fiber escapes digestion and absorption in the small intestine and becomes

Asian Crops and Human Dietetics

doi:10.1300/5504_01

available as a potential substrate for colonic microflora fermentation and provides antiglycemic and other potential disease-preventing benefits. Epidemiological studies have shown that a grain-based diet is associated with reduced risk of chronic diseases, including colorectal cancer (Jacobs et al. 1995), due to its phytochemical profile and antioxidant activities (Adom and Liu 2002; Adom et al. 2003; Charalampopoulos et al. 2002; Mori et al. 1999). The prevention of numerous chronic conditions by dietary fiber have been documented in epidemiological, case control, and clinical studies (Burkitt 1971; Howe et al. 1992; Jansen et al. 1991; Jacobs et al. 1998; Liu et al. 2000; Pereira et al. 2002; Slattery et al. 1997). In addition, it helps prevent obesity because high-fiber foods are reported as highly satiating (Blundell and Burley 1987), meaning that hunger is kept at bay longer after a high-fiber meal (Holt et al. 1995, 2001).

Animal tissues contain little carbohydrate relative to fat and protein, and hence cereal grains are the major sources of simple carbohydrates and dietary fiber. The food pyramid and other similar pictorial representations of dietary recommendations show this relative significance. Most dietary guidelines recommend that a major portion of a diet be comprised of grains. The importance of grains is reflected in the size of their segment in the U.S. food pyramid (Myfoodpyramid 2005) and in slogans like "Eat plenty of cereal foods, preferably whole grain," which can be found in several nations' dietary guidelines. Cereals are the edible seeds (fruits) of plants of the grass family and hence include the true millets by definition and description. Although the term *cereals* is used normally to refer to larger grains such as wheat and rice, which are the most widely consumed grains and sources of carbohydrates in the Western world, a number of small grains (such as millets) are popular in many parts of the world as sources of carbohydrates in the daily diets. Two of the major cereals produced and consumed today (rice and wheat) and a number of millets produced and consumed regularly in daily diets originated in Asia. In 2005 global cereal production was 2,227,980,030 Mt (FAOSTAT 2006). But in some developing countries, millet production exceeds cereal production. For example, in Sahelian Africa millet production is greater than cereal production where pearl millet (*Pennisetum glaucum*) production represents more than 75% of the total grain production (FAOSTAT 2006).

MAJOR CEREALS

Rice (Oryza sativa; *Poaceae)*

> Rice is a good choice for a healthy diet. It is a good source of complex carbohydrates and can help you achieve a reduced fat diet.
>
> American Dietetic Association

Rice is a food of such antiquity that the precise time and place of its origin may never be known; speculation has placed its origins in Africa, India, and parts of Southeast Asia. Rice was cultivated in India in approximately 3500 BC and in China in approximately 1500 BC. Archeological evidence points to rice as an important food in Mohanjodaro, India, as early as 2500 BC. Many world cultures consider rice a gift of the gods because of its long connection to human existence. Even today rice is treated with reverence, and its cultivation and processing are tied to important festivals and grand rituals in Asian cultures, especially those of the Hindus and Buddhists. For Indians, life begins and ends with the use of rice, and rice in celebrations, festivals, and funerals signifies important stages of growth and life events. For the Chinese, the most precious things are the "five grains," of which rice is the first. The Japanese believe that the emperor of Japan is the living embodiment of Ninigo-no-mikoto, the god of the ripened rice plant. For Thais, rice is a goddess, whose worship involves rituals for good growth and high yields and who is called on for reasons ranging from fortune-telling to requests for permission to use certain objects in nature.

Rice is the most cultivated and consumed food crop in Asia, where in many countries it covers almost half the cultivable land under production. In 2005 global rice production was 618,534,989 Mt and the area harvested was 153,783,818 ha (FAOSTAT 2006). Major rice-producing countries include China, India, Japan, Bangladesh, Indonesia, Thailand, Vietnam, Brazil, South Korea, the Philippines, and the United States. Cultivated rice varieties belong to two botanical species, *O. sativa* and *O. glaberrima,* of which *O. sativa* is more widely cultivated and consumed. Rice is an annual grass that can grow in paddies or marshland because its leaves transport oxygen to its roots. Rice plants grow to approximately 1.2 m (3.9 ft), and have long, narrow, flat leaves. The end of the leaf stalk bears the inflorescence (a panicle), with green and yellow flowers that bear the edible rice grains (fruits).

Of the more than 8,000 varieties of cultivated rice around the world, Indian basmati and Thailand jasmine rice have gained global acceptance and are perhaps the most renowned because of their distinct aroma and flavor. Rice is high in carbohydrate and has easily digestible proteins. Rice protein content ranges from 6.7% to 7.8%, carbohydrate from 77.2% to 81.7%, and fat from 0.5% to 2.9%. Rice processing yields three major kinds of rice grains currently available in retail markets: brown rice (only the husk is removed), white rice (husk, bran, and part of the germ are removed), and parboiled rice (rice with the husk is soaked and boiled before processing). Brown rice contains approximately 8% protein and traces of thiamine, niacin, riboflavin, iron, and calcium. White rice loses approximately 75% of its minerals during processing. Parboiled rice contains two to four times as much thiamine and niacin as raw white rice because the parboiling process causes these minerals to move from the outer bran to the center of the grain. Puffed and toasted rice is used as breakfast cereals; rice flour is used in cakes, pastries, and noodles; and fermented rice is used to make beer and liquors. Grain color (white, brown, red, and black), texture (grainy or sticky), scent, and flavor show great variability among the cultivated varieties.

Rice is a versatile ingredient in food-product development because it mixes well with a number of ingredients. The Chinese believe that rice strengthens the spleen and a weak stomach, increases appetite, and cures indigestion. In India a decoction of rice is prescribed as a salve to counteract an inflamed area of the skin. Rice water is valued as a traditional medicine for treating diarrhea. In folk medicine rice seeds are used for breast cancer, stomach indurations, and warts (Hartwell 1967-1971). Rice is also used as an astringent, diuretic, stomachic, tonic, and anthelmintic (vermifuge) and for abdominal ailments, burns, diarrhea, dysentery, dyspepsia, inflammation, nausea, psoriasis, skin ailments, sores, and swellings (Duke and Wain 1981).

Rice is consumed in its fermented form in many cultures. In southern India rice is fermented in several combinations with pulses and used to make breakfast and snack foods (e.g., *idli, dosa,* and *adai*).

When fermented with *Monascus* fungus it is referred to as Monascal rice (*anka*, red *koji*), a Chinese food for thousands of years. Monascal rice products are antioxidants (Yang et al. 2006) and have hypocholesterolemic (Endo and Monacolin 1979, 1980), hepatoprotective (liver-protective), and antitumor effects (Aniya et al. 1998; Yasukawa et al. 1996). Rice is also considered an essential part of wine making and is used in other fermented food products (Bau and Mo 1975).

Rice bran, a coproduct of milled rice, and its oil may have cardiovascular health benefits similar to those of oat bran (Most et al. 2005). However, humans consume little rice bran, primarily because of the rapid onset of rancidity in rice bran. But methods to produce stabilized rice bran (SRB) have been developed. SRB contains high-quality protein, oil, dietary fiber, polysaccharides, fat-soluble phytochemicals, and other bran nutrients. SRB is reported as a valuable nutraceutical product and an excellent dietary supplement for maintaining good health and preventing disease (Rukmini 2000). Studies have repeatedly shown that rice bran consists of almost 27% dietary fiber and has laxative and cholesterol-lowering abilities, thus making it an important additive and source of fiber in various functional foods (Hamid and Luan 2000). Rice bran oil is a healthy source of vegetable oil in Asian countries (Nicolosi et al. 1994; Sugano et al. 1999). Rice bran contains 10% to 23% oil (Saunders 1985) and larger amounts of insoluble dietary fiber. Several phytonutrients identified in rice bran have shown promising disease-preventing and health-related benefits, including prevention of cancers, hyperlipidemia, liver problems, hypercalciuria, kidney stones, and heart disease (Jariwalla 2001; Mori et al. 2000).

The nutritional values of long, medium, and short-grain rice within each variety or classification start out the same but vary with the way the variety is processed. Table 1.1 compares brown, unenriched white, and parboiled rice. Since populations consuming rice as a staple also tend to suffer from certain nutritional deficiencies, biofortification is being pursued to develop rice varieties richer in iron, zinc, and vitamin A (golden rice). Scientists have made significant headway with traditional breeding as well as modern biotechnological approaches. Recognizing the importance of rice and rice cultivation in human life and in poverty alleviation, several research centers and institutes have dedicated themselves to rice research. A good example is the International Rice Research Institute (IRRI) located in the Philippines, which

TABLE 1.1. Nutritional Value of Rice As Affected by Processing Method

Nutrient	Unit	Brown ¼ Cup Raw (46.25 g)	White (Unenriched) ¼ Cup Raw (46.25 g)	Parboiled (Unenriched) ¼ Cup Raw (46.25 g)
Proximates (macronutrients)				
Calories	kcal	171	169	172
Protein	g	3.64	3.30	3.14
Total fat	g	1.35	0.31	0.26
Carbohydrate	g	35.72	36.98	37.80
Fiber	g	1.62	0.60	0.79
Minerals				
Calcium, Ca	mg	10.64	12.95	27.75
Iron, Fe	mg	0.68	0.37	0.69
Magnesium, Mg	mg	66.14	11.56	14.34
Phosphorus, P	mg	154.01	53.19	62.90
Potassium, K	mg	103.14	53.19	55.50
Sodium, Na	mg	3.24	2.31	2.31
Zinc, Zn	mg	0.93	0.50	0.44
Copper, Cu	mg	0.13	0.10	0.09
Manganese, Mn	mg	1.73	0.50	0.39
Selenium, Se	mg	10.82	6.98	10.64
Vitamins				
Thiamine	mg	0.19	0.03	0.05
Riboflavin	mg	0.04	0.02	0.03
Niacin	mg	2.35	0.74	1.68
Pantothenic acid	mg	0.69	0.47	0.52
Vitamin B_6	mg	0.24	0.08	0.16
Folate	µg	9.25	3.70	7.86
Vitamin E	IU	0.33	0.06	0.07

Source: USDA. Nutrient Database for Standard Reference 2002.

focuses exclusively on rice as a major crop in the developing world. Good examples of IRRI's pathbreaking work include development of the high-yield IR-8 rice variety and a new strain of rice that has four to five times more iron than normal rice, since iron deficiency is the most common and prevalent micronutrient deficiency in the world. Please see www.irri.org periodically for updated work on rice research at this institute.

*Wheat (*Triticum *spp.; Poaceae)*

> Wheat flour is a good source of complex carbohydrate, the most efficient source of energy available to the human body.
>
> Minnesota Association of Wheat Growers

Wheat has been called the staff of life by humans and has been consumed by them since antiquity. The center of origin for wheat as given by Vavilov is southwest Asia. In 2005 global wheat production was 628,101,035 Mt and the area harvested was 215,614,524 ha. The top wheat producers are China, India, the United States, Russia, France, Canada, Australia, Germany, Pakistan, and Turkey (FAOSTAT 2006).

Common wheat, or bread wheat (*T. aestivum*), is the most widely cultivated wheat in the world. Other cultivated forms of wheat include durum (*T. durum*); einkorn (*T. monococcum*); emmer (*T. dicoccum*); spelt (*T. spelta*); and Kamut, or QK-77 (*T. polonicum* or *T. durum*).

Wheat is an annual grass with flat narrow leaves; erect, hollow, or pithy glabrous stems up to 1.5 m (5 ft) tall, bearing long, slender, somewhat compressed and flattened spikes. Each spike has two rows of spikelets, and each spikelet carries one to six florets. A head of wheat can contain 40 to 85 kernels.

Unlike rice, which is consumed as a whole grain and also as flour, wheat is usually consumed in preparations made from its flour, sometimes as fried or puffed grains, and rarely as whole grain. Wheat is ground after processing it into white wheat flour or whole-grain flour. One slice of whole wheat bread contains 1.5 g dietary fiber, while one slice of white bread contains only 0.5 g. Wheat is processed as cake flour, semolina, pastry flour, pasta flour, and self-rising flour. Prasad (1971) examined a number of wheat cultivars and found that wheat contained 8.9% to 15.8% protein, 2.60% to 1.89% ash, 0.321% to 0.483% phosphorus, 0.034% to 0.055% calcium, and 0.015% to 0.049% iron.

In India wheat has been used for treating diabetics since ancient times, and Ayurvedic literature of Indian traditional medicine calls it the "diabetic food." Along these lines, an increasing body of evidence suggests that foods with a low glycemic index (GI; carbohydrates that cause little change in blood glucose) are therapeutic as well as preventive for insulin resistance. Wheat has a low GI compared with rice (wheat flour 52 versus rice flour 65; Foster-Powell et al. 2002), and hence it is a suitable alternative for a diabetic diet. The chemical profile of wheat reveals the presence of a number of antioxidants (flavonoids, ferulic acid, carotenoids, etc.), which benefit health and have chemopreventive properties (Heim et al. 2002; Adom et al. 2003) by enhancing bowel health (McIntosh et al. 2003). The benefits of a whole wheat breakfast as a source of dietary antioxidants have been explored and reported in detail (Baublis et al. 2000). Wheat bran is used in functional food products because of its high-fiber content and antioxidant properties (Adom et al. 2003; Onyeneho and Hettiarachchy 1992; Zhao et al. 2004; Zhou et al. 2004). Wheat germ oil and bran oil are rich in tocopherols (vitamin E) and essential fatty acids (CSIR 1948-1976).

MILLETS

The survival of millets as cereals in competition with rice and wheat attests to the significance of millet crops in agroecosystems in Asia. *Millet* refers to a group of small cereal grains consumed as alternatives to the main staples rice and wheat. India is the largest producer of many kinds of millets, which are also called coarse cereals. They include six main grain crops: finger millet (*Eleusine coracana; ragi* in Tamil), kodo millet (*Paspalum scrobiculatum; varagu* in Tamil), little millet (*Panicum sumatrense; samai* in Tamil), foxtail millet (*Setaria italica; tenai* in Tamil), proso millet (*P. milliaceum; panivaragu* in Tamil), and barnyard millet (*Echinochloa frumentacea; kudiraivali* in Tamil). These millets have been cultivated for more than 5,000 years and are referred to very frequently in ancient Tamil literature. Most millets are dehulled and cooked in a way similar to rice cooking methods or powdered and used to make bread, thick porridge, and snacks. The popped, parboiled, and malted grains are also used regularly for preparation of sweet and savory dishes and snacks. Millets are also used in the preparation of flakes and noodles, whose preparation has become a cottage industry in India.

*Common Millet (***P. miliaceum;** *Poaceae)*

Common millet, also known as brown corn millet, broomcorn millet, Russian millet, proso millet, and hog millet, originated in Asia.

The plants grow to 6 ft and have hollow, pubescent stems. The inflorescence is clustered and compact, bearing round, elongated, or oval grains covered with a hull. The grains are red, pale yellow, or whitish gray. The dehulled grains are edible and are a common food in many parts of Asia. The nutritional values for millet are compared with rice in Table 1.2.

Millet bran is similar to rice and makes up about 15% of the weight of the grain. The bran contains 16% to 20% oil, 7.5% protein, 11.6% mineral elements, and 24% fiber and is a potential source of oil. Millet bran oil contains less oleic acid (26.0% versus 40%-50%) and more of the essential polyunsaturated fatty acid linoleic acid (52.0% versus 29%-42%) compared with rice bran oil (Jayachandra and Azeemoddin 1994). Beer made from millet is a source of linoleic acid.

*Finger Millet (*Eleusine coracana; *Poaceae)*

Finger millet is an important cereal in India, Africa, and Sri Lanka. Finger millet ranks fourth in production, after wheat, rice, and maize, in India. The grain is higher in protein, fat, and minerals than rice, corn, or sorghum (Reed 1976). It is usually converted into flour and made into cakes, puddings, or porridge. Finger millet has been in routine use in Asian cultures for hundreds of years. It is consumed as *muddae* (a thick porridge), *dosa* (a pancake), biscuits, and malt by many southern Indians, especially in the states of Tamil Nadu, Andhra Pradesh, and Karnataka (Rao and Muralikrishna 2001). A fermented drink or beer is also made from the grain. The grain can be malted (softened and germinated) and a flour of the malted grain used as a nourishing food for infants and invalids.

Finger millet leaf juice has been given to women in childbirth, and the plant parts are diaphoretic and diuretic; an anthelmintic; and a remedy for leprosy, liver disease (Watt and Breyer-Brandwijk 1962), measles, pleurisy, pneumonia, and smallpox (Duke and Wain 1981).

Malted finger millet's acclaimed functional food status results from it being a good source of fiber and nutrients. Extensive studies have been carried out on changes taking place during the malting of finger millet, including research on (1) carbohydrate-degrading enzymes and characterization of alpha-amylases (Nirmala et al. 2000; Nirmala and Muralikrishna 2003); (2) nonstarch polysaccharides and their func-

TABLE 1.2. Nutrient Composition of Millets, Rice, Wheat, and Other Cereals

Grain	Protein (g)	Fat (g)	Crude Fiber (g)	Carbohydrate (g)	Ca (mg)	Fe (mg)	Thiamine (mg)	Riboflavin (mg)	Niacin (mg)
Rice (brown)	7.9	2.7	1.0	76.0	33	1.8	0.41	0.04	4.3
Wheat	11.6	2.0	2.0	71.0	30	3.5	0.41	0.10	5.1
Finger millet	7.7	1.5	3.6	72.6	350	3.9	0.42	0.19	1.1
Foxtail millet	11.2	4.0	6.7	63.2	31	2.8	0.59	0.11	3.2
Common millet	12.5	3.5	5.2	63.8	8	2.9	0.41	0.28	4.5
Little millet	9.7	5.2	7.6	60.9	17	9.3	0.30	0.09	3.2
Barnyard millet	11.0	3.9	13.6	55.0	22	18.6	0.33	0.10	4.2
Kodo millet	9.8	3.6	5.2	66.6	35	1.7	0.15	0.09	2.0

Source: Hulse et al. 1980; United States National Research Council/National Academy of Sciences 1982; United States Department of Agriculture/Human Nutrition Information Service 1984.

Note: Values are Per 100 g edible portion; 12% moisture.

tional properties (Rao et al. 2004); (3) phenolic compounds and their antioxidant properties (Rao and Muralikrishna 2002); and (4) free sugars, amino acids, phytic acid, minerals, and dietary fiber content (Antony et al. 1996; Malleshi et al. 1986; Nirmala et al. 2000). In addition, several beverages and pharmaceutical preparations have been formulated by using germinated finger millet flour (Mwikya et al. 2002).

A nutritional analysis of finger millet is given in Table 1.2. It is considered a good alternative to rice and a wholesome food for diabetic patients. Finger millet has also come to be known for its excellent hypoglycemic and hypocholesterolemic characteristics (Gopalan 1981; Gopalan et al. 1989; Madhusudhan and Tharanathan 1995) and antioxidant properties (Sripriya et al. 1996). Finger millet is reported to be a richer source of dietary fiber (18%), phenolic compounds (1%), and calcium (0.34%) than cereals such as barley, rice, maize, and wheat (Ravindran 1991). Other studies have shown that ferulic acid present in finger millet is a potential chemopreventive agent for colorectal cancer (Kawabata et al. 2000; Mori et al. 1999). Colorectal cancer is a major cause for concern in developed and developing nations' health programs; its five-year survival rate is as low as 60%.

Finger millet was once considered a poor person's food, mostly consumed by people engaged in hard labor and by the poor and not included in the diets of the urban and the elite. Of late, with realization of finger millet's nutritional value and its important role in human health promotion, its use has increased across all populations in India. It is mixed proportionately with other grains and grain products to prepare savory and sweet dishes.

*Kodo Millet (***Paspalum scrobiculatum;** *Poaceae)*

Kodo millet is of high protein and organoleptic quality (Kulkarni and Naik 2000) and a rich source of phenolics, tannins, and phytates, which can act as antioxidants (Hegde and Chandra 2005; Hegde et al. 2005). Some reports suggest that it has more dietary fiber (9 g versus 3.6 g per 100 g of edible portion) and iron (12 mg versus 9 mg per 100 g of editble portion) (Senthil et al. 2005) than finger millet. A nutritional analysis of kodo millet is given in Table 1.2.

A comparison of the nutritive values of kodo millet and finger millet with that of rice and wheat is given in Table 1.3.

TABLE 1.3. Nutrional Values of Kodo Millet, Finger Millet, Rice, and Whole Wheat

Grain	Protein (g)	Carbohydrate (g)	Fat (g)	Total Fiber (g)	Energy (J)	Phytin (mg)	Ca (mg)	Fe (mg)	Zn (mg)	P (mg)
Kodo millet	8.3	65.9	1.4	37.8	1,293.00	135	27	0.5	0.7	188
Finger millet	7.3	72.0	1.3	19.1	1,372.35	209	344	3.9	2.3	283
Milled rice	6.8	78.2	0.5	NA	1,443.48	83	10	0.7	1.3	160
Whole wheat	11.8	71.2	1.5	11.4	1,447.66	238	41	5.3	2.7	306

Source: Hedge and Chandra 2005.

Note: Values are per 100 g edible portion; NA = not applicable.

*Foxtail Millet (*Setaria italica; *Poaceae)*

Although foxtail millet is known widely as Italian, Hungarian, Siberian, or German millet, Vavilov suggested its origin as eastern Asia (China) and India. Foxtail millet was the most important plant food in Neolithic culture in China and parts of India, and its domestication and cultivation were the earliest identifiable manifestations of these cultures at more than 4,000 to 5,000 years ago (Chang 1968). It is described in Tamil literature as the favored millet for the Tamil god Lord Muruga and is offered during religious festivals at temples and other places of worship. Foxtail millet ranks second in total world production of millets and has an important place in world agriculture, especially in subtropical and tropical Asia (Marathee 1993). For its nutritional composition see Table 1.2. Foxtail millet's protein improves insulin sensitivity and cholesterol metabolism and has been suggested as a beneficial food component in obesity-related diseases such as type 2 diabetes and cardiovascular diseases (Choi et al. 2005).

*Barnyard Millet (*Echinochloa frumentacea; *Poaceae)*

Barnyard millet originated in Japan and other parts of Asia and is one of the fastest growing minor millets. *E. frumentacea* is more commonly grown in India, China, Korea, and Japan. A nutritional analysis of barnyard millet is given in Table 1.2. Veena et al. (2005) analyzed nine cultivars of barnyard millet grown in India and determined moisture (8.1%-9.8%), protein (8.4%-15.5%), fat (3.2%-4.5%), total minerals (2.7%-4.2%), calcium (17.1-32.7 mg/100 g), iron (1.2-1.5 mg/100 g), starch (51.5%-59.5%), dietary fiber (soluble 3.5%-4.6% and insoluble 6.1%-10.5%), and in vitro digestibility of protein (74.9%-84.7%). Use of barnyard millet for organoleptic and nutritional enhancement of traditional rice-based Indian dishes (*idli, dosa, roti, chakli*) is being explored.

Japanese Barnyard Millet
*(*Panicum frumentaceum; *Poaceae)*

More commonly grown and consumed in Japan, Japanese barnyard millet is a highly nutritious cereal that is rich in protein, lipids, vitamins

B_1 and B_2, and nicotinic acid compared with other cereals, such as rice and wheat grains (Resources Council, Science and Technology Agency, Japan 1982). Taira (1983) described the amino acid composition of its protein as "good" and oleic and linoleic acids as "abundant," while Watanabe (1999) demonstrated the presence of antioxidant phenolic compounds. Recently, Japanese barnyard millet grains have been fed, in place of rice and wheat grains, to patients with allergic diseases including atopic dermatitis (Urisu et al. 1991).

*Little Millet (***P. sumatrense, P. miliare;** ***Poaceae)***

Little millet originated in Southeast Asia and is called *samai* in Tamil. It is an annual erect plant growing to 0.9 to 1.2 m (3 to 4 ft), whose long glabrous leaf blades have acuminate tips. The inflorescence is a panicle, 10 to 40 cm long and 2 to 8 cm wide. It is widely distributed in tropical and subtropical Asia (India, eastern Asia, Indochina, and Malaysia). It is a popular millet in villages and grown mostly for local consumption. For its nutritional composition see Table 1.2.

BUCKWHEAT

Buckwheat is an edible grain belonging to the grass family and native to northern Asia. Buckwheat is not a true cereal and is often called a pseudocereal because its seeds are used as flour for bread and other staples. It is cultivated in China, the United States, Canada, Japan, and other parts of the world.

*Buckwheat (***Fagopyrum esculentum;** ***Polygonaceae)***

Buckwheat is an ideal crop to cultivate because of its hardiness and short growing season (70-90 days). Buckwheat plants are 0.6 to 0.9 m (2 to 3 ft) tall and grow rapidly, bearing small heart-shaped, broad leaves with slender, hollow stems. The brown seeds are as large as soybeans, irregularly shaped, and have four triangular surfaces. The seeds are dehulled, and the remaining seed material, called groat, is ground into flour. The flour is often mixed with flour from other cereal grains to make breads, breakfast cereals, or other multigrain products. In Japan buckwheat flour is mixed with wheat flour to make soba noodles. In

Russia it is used in a variety of foods, including roasted whole groats to make kasha. Buckwheat is high in lysine and so it can be added to low-lysine wheat and corn to make a more nutritive mixture.

The nutritional value of buckwheat per 100 g sample is reported as fat, 2.4 g; protein, 14.4 g; dietary fiber, 8.5 g; niacin, 4.9 mg; thiamine, 0.46 mg; riboflavin, 0.14 mg; vitamin B_6, 0.73 mg; pantothenic acid, 1.05 mg; carotenoids, 0.21 mg; tocopherols, 0.66 mg; potassium, 455 mg; calcium, 20 mg; magnesium, 206 mg; phosphorus, 472 mg; zinc, 2.8 mg; iron, 2.3 mg; and manganese, 1.7 mg (Gabrovska et al. 2002). Buckwheat recently has been recognized as a healthy food and its place in the human diet is being reexamined because of its high content of vitamins B_1 and B_2, the high biological value of its protein rich in lysine (Pomeranz and Robbins 1972; Sure 1955), and its demonstrated hypotensive effect (Matsubara et al. 1985). Demand for buckwheat grain is steadily improving, as populations seek healthy foods for functionality, health promotion, and disease prevention. Demand for this grain has increased because of growing popularity of multigrain baked foods, which have more nutritional value and flavor.

It has also been suggested that this pseudocereal influences appetite because of its high satiating efficiency index (SEI) compared with traditional cereal foods (Berti et al. 2005). Foods with a high SEI delay the return of hunger and hence have the potential to reduce obesity. Antioxidant activity of buckwheat seedlings, seeds, hulls, groats, and grains, due to phenolic compounds including flavonoids, has been reported by several researchers (Luthar and Kreft 1996; Margna and Margna 1982; Sato and Sakamura 1975; Watanabe et al. 1995; Watanabe 1998), suggesting the further value of this plant food.

REFERENCES

Adom KK, Liu RH. 2002. Antioxidant activity of grains. Journal of Agricultural and Food Chemistry 50: 6182-6187.

Adom KK, Liu RH, Sorrells ME. 2003. Phytochemical profiles and antioxidant activity of wheat varieties. Journal of Agricultural and Food Chemistry 51: 7825-7834.

Aniya Y, Yokomakura T, Yonamine M, Nagamine T, Nakanishi H. 1998. Protective effect of the mold *Monascus anka* against acetaminophen-induced liver toxicity in rats. Japanese Journal of Pharmacology 78: 79-82.

Antony U, Sripriya G, Chandra TS. 1996. Effect of fermentation of the primary nutrients in finger millet (*Eleusine coracana*). Journal of Agricultural and Food Chemistry 44: 2616-2618.

Bau YS, Mo CF. 1975. The uses and culturing methods of *Monascus purpureus* Went. New Asia College Academy Annual 17: 335-357.

Baublis AJ, Lu C, Clydesdale FM, Decker EA. 2000. Potential of wheat-based breakfast cereals as a source of dietary antioxidants. Journal of the American College of Nutrition 19(3): 308S-311S.

Berti C, Riso P, Brusamolino A, Porrini M. 2005. Effect on appetite control of minor cereal and pseudocereal products. British Journal of Nutrition 94(5): 850-858.

Blundell JE, Burley VJ. 1987. Satiation, satiety and the action of fibre of food intake. International Journal of Obesity 11(S): S9-S25.

Burkitt DP. 1971. Epidemiology of cancer of the colon and rectum. Cancer 28: 3-13.

Chang K. 1968. Archeology of ancient China. Science 162: 519-526.

Charalampopoulos D, Wang R, Pandiella SS, Webb C. 2002. Application of cereals and cereal components in functional foods: A review. International Journal of Food Microbiology 79: 131-141.

Choi YY, Osada K, Yoshiaki ITO, Nagasawa T, Choi MR, Nishizawa N. 2005. Effects of dietary protein of Korean foxtail millet on plasma adiponectin, HDL-cholesterol, and insulin levels in genetically type 2 diabetic mice. Bioscience, Biotechnology, and Biochemistry 69: 31-37.

CSIR (Council of Scientific and Industrial Research). 1948-1976. The wealth of India. 11 vols. CSIR, New Delhi, India.

Duke JA, Wain KK. 1981. Medicinal plants of the world. A computer index with more than 85,000 entries. 3 vols. Food and Agricultural Organization, United Nations, Rome.

Endo A, Monacolin K. 1979. A new hypocholesterolemic agent produced by a *Monascus* species. Journal of Antibiotics 32: 852-854.

Endo A, Monacolin K. 1980. A new hypocholesterolemic agent that specifically inhibits 3-hydroxy-3-methylglutaryl coenzyme A reductase. Journal of Antibiotics 33: 334-336.

FAOSTAT (Statistics Division of the Food and Agriculture Organization of the United Nations). 2006. http://faostat.fao.org/. Accessed June 20, 2006.

Foster-Powell K, Holt SHA, Brand-Miller JC. 2002. International table of glycemic index and glycemic load values, 2002. American Journal of Clinical Nutrition 76(1): 5-56.

Gabrovska D, Fiedlerova V, Holasova M, Maskova E. 2002. The nutritional evaluation of underutilized cereals and buckwheat. Food and Nutrition Bulletin 23(3): 246-249.

Gopalan C. 1981. Carbohydrates in diabetic diet. Bulletin of Nutrition Foundation, India, July. p. 3.

Gopalan C, Ramashastri BV, Balasubramanian BV. 1989. Nutritive value of Indian foods. ICMR Offset Press, New Delhi, India. p. 47-91.

Hamid AA, Luan YS. 2000. Functional properties of dietary fiber prepared from defatted rice bran. Food Chemistry 68(1): 15-19.

Hartwell JL. 1967-1971. Plants used against cancer. A survey. Lloydia: 30-34.

Hegde PS, Chandra TS. 2005. ESR spectroscopic study reveals higher free radical quenching potential in kodo millet (*Paspalum scrobiculatum*) compared to other millets. Food Chemistry 92: 177-182.

Hegde PS, Rajasekaran NS, Chandra TS. 2005. Effects of the antioxidant properties of millet species on oxidative stress and glycemic status in alloxan-induced rats. Nutrition Research 25(12): 1109-1120.

Heim KE, Anthony RT, Dennis JB. 2002. Flavonoid antioxidants: Chemistry, metabolism and structure-activity relationships. Journal of Nutritional Biochemistry 13: 572-584.

Holt SHA, Brand-Miller JC, Petocz P, Farmakalidis E. 1995. A satiety index of common foods. European Journal of Clinical Nutrition 49: 675-690.

Holt SHA, Brand-Miller JC, Stitt PA. 2001. The effects of equal-energy portions of different breads on blood glucose levels, feelings of fullness and subsequent food intake. Journal of the American Dietetic Association 101(7): 767-773.

Howe GR, Benito E, Castelleto R, Cornee J, Esteve J, et al. 1992. Dietary intake of fiber and decreased risk of cancers of the colon and rectum. Evidence from the combined analysis of 13 case-control studies. Journal of the National Cancer Institute 84: 1887-1896.

Hulse, JH, Laing EM, Pearson OK. 1980. Sorghum and the millets: Their compositions and nutritive value. Academic Press, New York.

Jacobs D, Meyer K, Kushi L, Folsom A. 1998. Whole-grain intake may reduce the risk of ischemic heart disease death in post-menopausal women: The Iowa women's health study. American Journal of Clinical Nutrition 68: 248-257.

Jacobs DR, Slavin J, Marquart L. 1995. Whole grain intake and cancer: A review of literature. Nutrition and Cancer 22: 221-229.

Jansen MCJF, Bueno-de-Mesquita HB, Buzina R, Buzina R, Fidanza F, et al. 1991. For the Seven Countries Study Research Group. Dietary fiber and plant foods in relation to colorectal cancer mortality: The seven countries study. International Journal of Cancer 81: 174-179.

Jariwalla RJ. 2001. Rice-bran products: Phytonutrients with potential applications in preventive and clinical medicine. Drugs Under Experimental and Clinical Research 27: 17-26.

Jayachandra K, Azeemoddin G. 1994. Common millet (*Panicum miliaceum*) bran: A potential source of oil. Journal of the Oil Technologists' Association of India 26(3): 73-74.

Kawabata K, Yamamoto T, Hara A, Shimizu M, Yamada Y, et al. 2000. Modifying effects of ferulic acid on azoxymethane-induced colon carcinogenesis in F344 rats. Cancer Letters 157: 15-21.

Kulkarni LR, Naik RK. 2000. Nutritive value, protein quality and organoleptic quality of kodo millet (*Paspalum scrobiculatum*). Karnataka Journal of Agricultural Sciences 13(1): 125-129.

Liu S, Manson J, Stampfer M, Hu F, Giovannucci E, et al. 2000. A prospective study of whole-grain intake and risk of type 2 diabetes mellitus in US women. American Journal of Public Health 90: 1409-1415.

Luthar Z, Kreft I. 1996. Composition of tannin in buckwheat (*Fagopyrum esculentum* Moench) seeds. Research Report BFUL 67: 59-65.

Madhusudhan B, Tharanathan RN. 1995. Legume and cereal starches: Why differences in digestibility? Part II. Isolation and characterization of starches from rice (*O. sativa*) and ragi (*E. coracana*). Carbohydrate Polymers 28: 153-158.

Malleshi NG, Desikachar HSR, Tharanathan RN. 1986. Free sugars and non-starch polysaccharides of finger millet (*Eleusine coracana*), pearl millet (*Pennisetum typhoideum*), foxtail millet (*Setaria italica*) and their malts. Food Chemistry 20: 253-261.

Marathee JP. 1993. Structure and characteristics of the world millet economy. In: Riley KW, Gupta SC, Seetharam A, Mushonga JN, editors. Advances in small millets. Oxford and IBH Publishers, New Delhi, India. p. 159-178.

Margna UV, Margna ÉR. 1982. Differential nature of quantitative shifts in flavonoid accumulation in buckwheat seedlings of different ages. Soviet Plant Physiology 29: 223-230.

Matsubara Y, Kumamoto H, Iizuka Y, Murakami T, Okamoto K, et al. 1985. Structure and hypotensive effect of flavonoid glycosides in *Citrus unshiu* peelings. Agricultural and Biological Chemistry 49: 909-914.

McIntosh GH, Noakes M, Royle PJ, Foster PR. 2003. Whole-grain rye and wheat foods and markers of bowel health in overweight middle-aged men. American Journal of Clinical Nutrition 77(4): 967-974.

Mori H, Kawabata K, Matsunaga K, Ushida J, Fujii K, et al. 2000. Chemopreventive effects of coffee bean and rice constituents on colorectal carcinogenesis. Biofactors 12: 101-105.

Mori H, Kawabata K, Yoshimi N. 1999. Chemopreventive effects of ferulic acid and rice germ on large bowel carcinogenesis. Anticancer Research 19: 3775-3779.

Most MM, Tulley R, Morales S, Lefevre M. 2005. Rice bran oil, not fiber, lowers cholesterol in humans. American Journal of Clinical Nutrition 81(1): 64-68.

Mwikya SM, Camp JV, Mamiro PRS, Ooghe W, Kolsteren P, Huyghebaert A. 2002. Evaluation of the nutritional characteristics of a finger millet based complementary food. Journal of Agricultural and Food Chemistry 50: 3030-3036.

Myfoodpyramid. 2005. http://www.mypyramid.gov/. Accessed July 18, 2006.

Nicolosi RJ, Rogers E, Ausman LM, Orthoefer FT. 1994. Rice bran oil and its health benefits. Food Science and Technology 59: 421-437.

Nirmala M, Muralikrishna G. 2003. Three α-amylases from malted finger millet (ragi, *Eleusine coracana*, Indaf-15): Purification and partial characterization. Phytochemistry 62: 21-30.

Nirmala M, Rao SMVSST, Muralikrishna G. 2000. Carbohydrates and their degrading enzymes from native and malted finger millet (ragi, *Eleusine coracana*, Indaf-15). Food Chemistry 69: 175-180.

Onyeneho SN, Hettiarachchy NS. 1992. Antioxidant activity of durum wheat bran. Journal of Agricultural and Food Chemistry 40: 1496-1500.

Pereira M, Jacobs D, Pins J, Raatz S, Gross M, et al. 2002. Effect of whole grains on insulin sensitivity in overweight hyperinsulinemic adults. American Journal of Clinical Nutrition 75: 848-855.

Pomeranz Y, Robbins GS. 1972. Amino acid composition of buckwheat. Journal of Agricultural and Food Chemistry 20: 270-274.

Prasad B. 1971. Nutritional value of wheat. Indian Journal of Chemistry 43(4): 145-147.

Rao SMVSST, Muralikrishna G. 2001. Non-starch polysaccharides and bound phenolic acids. Food Chemistry 72(2): 187-192.

Rao SMVSST, Muralikrishna G. 2002. Evaluation of the antioxidant properties of free and bound phenolic acids from native and malted finger millet (ragi, *Eleusine coracana*, Indaf-15). Journal of Agricultural and Food Chemistry 50: 889-892.

Rao SMVSST, Muralikrishna G, Manohar SR. 2004. Functional characteristics of non-starch polysaccharides (NSP) obtained from native (n) and malted (m) finger millet (ragi, *Eleusine coracana*, Indaf-15). Food Chemistry 88: 453-460.

Ravindran G. 1991. Studies on millets: Proximate composition, mineral composition, phytate and oxalate contents. Food Chemistry 39: 99-107.

Reed CF. 1976. Information Summaries on 1000 Economic Plants. U.S. Department of Agriculture, Washington, DC.

Resources Council, Science and Technology Agency. 1982. Standard Tables of Food Composition. 4th ed. Resources Council, Science and Technology Agency, Japan. p. 707.

Rukmini C. 2000. Bioactive in rice bran and rice bran oil. In: Bidlack WR, Omaye ST, Meskin MS, Topham DKW, editors. Phytochemicals as bioactive agents. Technomic, Lancaster, PA. p. 213-240.

Sato H, Sakamura S. 1975. Isolation and identification of flavonoids in immature buckwheat seed (*Fagopyrum esculentum* Mönch). Journal of the Agricultural Chemical Society of Japan 49: 53-55.

Saunders RM. 1985. Rice bran: Composition and potential food uses. Food Reviews International 1: 465-495.

Senthil A, Nirmala KA, Joel JA, Selvi B, Raveendran TS, et al. 2005. Small millets for nutritional security. Kalaiselvam Pathipagam, Tamil Nadu Coimbatore, India. p. 1-11.

Slattery ML, Potter JD, Coates A, Ma KN, Berry TD, et al. 1997. Plant foods and colon cancer: An assessment of specific foods and their related nutrients (United States). Cancer Causes and Control 8: 575-590.

Sripriya G, Chandrasekharan K, Murthy VS, Chandra TS. 1996. ESR spectroscopic studies on free radical quenching action of finger millet (*Eleusine coracana*). Food Chemistry 57: 537-540.

Sugano M, Koba K, Tsuji E. 1999. Health benefits of rice bran oil. Anticancer Research 19: 3651-3657.

Sure B. 1955. Nutritive value of proteins in buckwheat and their role as supplements to proteins in cereal grains. Journal of Agricultural and Food Chemistry 3: 793-795.

Taira, H. 1983. Influence of flooded culture on lipid content and fatty acid composition of Japanese barnyard millet. Report of the National Food Research Institute 43: 54-57.

Urisu A, Yamada K, Masuda S, Komada H, Wada E, et al. 1991. 16-kd rice protein is one of the major allergens in rice grain extract and responsible for crossallergenicty between cereal grains in Poacea family. International Archives of Allergy and Applied Immunology 96: 244-252.

United States Department of Agriculture/Human Nutrition Information Service (USDA/HNIS). 1984. Composition of Foods: Cereal Grains and Pasta. Agriculture Handbook No. 8-20. U.S. Department of Agriculture, Washington, DC.

United States National Research Council/National Academy of Sciences. 1982. United States Canadian Tables of Food Composition. 3rd ed. rev. National Academy Press, Washington, DC.

USDA Nutrient Database for Standard Reference. 2002. Nutrient Composition Database for Standard Reference, Release 15 (August). U.S. Department of Agriculture, Washington, DC.

Veena B, Chimmad BV, Naik RK, Shanthakumar G. 2005. Physico-chemical and nutritional studies in barnyard millet. Karnataka Journal of Agricultural Sciences 18(1): 101-105.

Watanabe M. 1998. Catechins as antioxidants from buckwheat (*Fagopyrum esculentum* Moench) groats. Journal of Agricultural and Food Chemistry 46: 839-845.

Watanabe M. 1999. Antioxidative phenolic compounds from Japanese barnyard millet (*Echinochloa utilis*) grains. Journal of Agricultural and Food Chemistry 47(11): 4500-4505.

Watanabe M, Sato A, Osawa R, Terao J. 1995. Antioxidative activity of buckwheat seed extracts and its rapid estimate for evaluation of breeding materials. Journal of the Japanese Society for Food Science and Technology 42: 649-655.

Watt JM, Breyer-Brandwijk MG. 1962. The medicinal and poisonous plants of southern and eastern Africa. 2nd ed. E and S Livingstone, Edinburgh.

Yang JH, Tseng YH, Lee YL, Mau JL. 2006. Antioxidant properties of methanolic extracts from monascal rice. LWT—Food Science and Technology 39(7): 740-747.

Yasukawa K, Takahashi M, Yamanouchi S, Takido M. 1996. Inhibitory effect of oral administration of *Monascus* pigment on tumor promotion in two-stage carcinogenesis in mouse skin. Oncology 53: 247-249.

Zhao C, Stini WA, Marshall JR, Martinez ME, Rodriguez JMG, et al. 2004. Wheat bran fiber supplementation and bone loss among older people. Nutrition 20: 747-751.

Zhou K, Su L, Yu LL. 2004. Phytochemicals and antioxidant properties in wheat bran. Journal of Agricultural and Food Chemistry 52: 6108-6114.

Chapter 2

Vegetables

Worldwide, national food and dietary guidelines recommending increased consumption of dark leafy and other colorful vegetables are based on accumulating evidence that vegetable consumption is associated with a lower risk of coronary heart disease, diabetes, and cancer. Total global vegetable production increased from 564,451,746 Mt in 1995 to 883,144,797 Mt in 2005 (FAOSTAT 2006). The health benefits of this group of plant foods are due to their phytochemicals, nutrient compounds, and anutrient compounds, which are most often produced as plant defense mechanisms to overcome extremes of environmental conditions or pest and pathogenic attacks. Phytochemicals have been studied extensively by plant scientists and are now being studied by dietitians for consideration in dietary plans and recommendations.

PHYTOCHEMICALS, NUTRIENT AND ANUTRIENT COMPOUNDS

Vitamins

Although the importance of vitamins in the human diet has long been known, and they are included in dietary plans mainly to alleviate distinct deficiency symptoms and diseases, they are receiving greater attention lately because of their antioxidants, which are necessary in the prevention of degenerative diseases. In general, antioxidant vitamins, including ascorbic acid and precursors of vitamins E and A, protect body tissues against the damaging effects of free oxygen radicals

Asian Crops and Human Dietetics

doi:10.1300/5504_02

produced during routine metabolic functions. There is strong evidence that tocopherols (vitamin E) play an important role in preventing chronic diseases such as heart disease and several types of cancer by inhibiting the oxidation of polyunsaturated fatty acids in plasma lipoproteins. Vitamin E protects against approximately 80 diseases, including cancer and cardiovascular diseases (Gey and Puska 1989); cell membrane and DNA damage by free radicals; oxidation of low-density lipoproteins; and disorders of the skin, eye, and lungs (Agudo et al. 1997; Kurilich et al. 1999; Machlin 1984; Moushumi and Sumati 1994; Patterson et al. 1997; Verreault et al. 1989; Watson and Leonard 1986; Weber et al. 1997). The most biologically active is alpha-tocopherol, the most common form of vitamin E present in nature (Bjorneboe et al. 1990), which is preferentially retained in large quantities and transported to body components (Traber et al. 1990). The current dietary reference intake (DRI) for vitamin E is 15 mg for adults (National Academies 2001), compared with the previous recommended dietary allowance (RDA) of 10 mg for male adults and 7.6 mg for female adults (National Research Council 1989).

Glucosinolates and Isothiocyanates

Glucosinolates, largely present in cruciferous vegetables, and their breakdown products, such as isothiocyanates, have demonstrated protection against various types of cancers (Fahey et al. 2002; Kostova et al. 2005; Zhang and Talalay 1994). The antioxidant activities of sulforaphane, phenethyl isothiocyanate, indole-3-carbinol, glucaric acid, and other isothiocyanates offer protection and detoxification of carcinogens at multiple levels of defense. Some studies have suggested that these compounds lower risk of atherosclerosis and cancer (Hecht 1999).

Phytosterols

Phytosterols are phytochemicals so named because of their close resemblance to cholesterol. The plant sterols differ from cholesterol by having ethyl or methyl groups or unsaturation in the side chain. The predominant sterols include sitosterol, stigmasterol, and campesterol, and are found in foods such as wheat germ, soybeans, and corn oil. Numerous health benefits of phytosterols such as reduction of cholesterol levels, prevention of many diseases and cancer (colon, prostate, and breast), inhibition of tumor growth, and stimulation of apoptosis have been reported (Awad et al. 2000; Ovesna et al. 2004).

Polyphenols

Previously considered to be anutrient compounds causing toxicity and as such undesired in the human diet, polyphenols are currently being reexamined because of many testimonies of their efficiency as antioxidants stronger than the antioxidant vitamins (Bravo 1998; Shrikhande and Franch 1974). Simple phenols, polyphenols, and tannins have been of great interest in the past because of their effect on the color, odor, and flavor of foods and beverages (Shahidi and Naczk 1995). Flavonoids with diphenylpropane ($C_6C_3C_6$) skeletons are the most common polyphenols in fruits and vegetables (Bravo 1998) and red wine (Vinson et al. 1995). This group of compounds consists of as many as 4,000 secondary plant metabolites, many present as sugar conjugates. Epidemiological studies have related diets rich in flavonols to a reduced incidence of heart disease (Hertog 1995). Świderski (2000) suggested a daily consumption of "not less than 1 to 2 g" polyphenols (including flavonoids). They have gained greater importance recently because of their described health-protecting properties (Frei and Higdon 2003; Sampson et al. 2002; Sesso et al. 2003; Stewart et al. 2003). Flavonoids demonstrate a wide range of biochemical and pharmacological effects, including antioxidant, anti-inflammation, antiplatelet, antithrombotic, and antiallergic effects. The mechanisms of activity of flavonoids have been studied widely: (1) inhibition of enzymes related to tumorigenesis such as prostaglandin synthase, lypoxygenase, and cyclooxygenase (Baumann et al. 1980; Laughton et al. 1991); (2) induction of detoxifying enzymes such as glutathione S-transferase (Smith and Yang 1994); (3) inhibition of oxidation and cytotoxicity of low-density lipoprotein by quercetin (De Whalley et al. 1990); and (4) reduction of coronary heart disease and cancer risks (Yoshida et al. 1990). Vinson et al. (1995) demonstrated that the flavonoids quercetin, myricetin, and rutin are more powerful antioxidants in vitro than the traditional vitamins. Flavonols and flavones are described as possessing antioxidant and free radical scavenging activity (Hollman et al. 1997) and "vitamin C sparing activity" (Middleton and Kandaswami 1992). Myricetin also exhibits vitamin C sparing activity and is described as the flavonoid having the most activity of this type (Larson 1988). Myricetin is a good antioxidant and a potent anticarcinogen and antimutagen, preventing atherosclerosis. Quercetin is an antimutagenic that inhibits human prostate and lung cancer cell growth (Le-Marchand et al. 2000; Xing et al. 2001) and reduces the incidence of cardiovascular diseases (Knekt et al. 2000). It is a chemopreventive and chemotherapeutic agent that can relieve local pain caused by inflammation, headache, oral surgery, and stomach ulcer. Some polyphenols are described as phytoestrogens because of their estrogenic or antiandrogenic activity, or both. These include isoflavones, stilbenes, lignans, and coumarins (Cassidy et al. 2000). However, lignans are not estrogenic until they are transformed by the gut microflora (Cassidy et al. 2000; Setchell et al. 1981). Another flavonoid being studied for its health benefits is kaempferol, which has antioxidant, antitumor, anti-inflammatory, and antiulcerogenic activity.

Plant Pigments

The importance of plant pigments in photosynthesis and plant antioxidant activity has long been known. Their value to the human diet is demonstrated in dietary recommendations to choose a variety of colors. Many plant pigments are correlated with a specific human body function or part. Lutein and zeaxanthin have been associated with lower risk of developing cardiovascular disease (Granado et al. 2003), several types of cancers (Granado et al. 2003), cataracts (Olmedilla et al. 2003), and age-related macular degeneration (Krinsky et al. 2003). Lutein and zeaxanthin are the dominant carotenoids in the human retina (Sommerburg et al. 1999), representing about 36% and 18%, respectively, of the total carotenoid content of the retina (Landrum and Bone 2001). They are also the only carotenoids that can be detected in the human lens (Yeum et al. 1995) because they are selectively deposited from the blood into the macular region (Chan et al. 1998), a region that is rich in cone receptors (Krinsky et al. 2003). These pigments' antioxidant role (Dagnelie et al. 2000; Semba and Dagnelie 2003) and their absorption of the most energetic portion of visible light, which causes the most damage (Krinsky et al. 2003), have been demonstrated. Lutein and zeaxanthin cannot be synthesized by humans and must be obtained through the diet (Semba and Dagnelie 2003). Although high intake of lutein has been shown to have potential health benefits, it is not considered an essential nutrient (Harper 1999) but a conditionally essential nutrient (Semba and Dagnelie 2003). Other pigments studied for their health benefits include chlorophyll, which helps prevent colon cancer (de Vogel et al. 2005), and lycopene, a red pigment present in tomato, watermelon, and other red vegetables that has been associated strongly with prevention of prostate cancer. The dietary guidelines by the Produce for Better Health Foundation (2007) clearly connect plant color to a phytochemical constituent in fruits and vegetables and list their health benefits (see Table 2.1).

CUCURBITS

Cucurbitaceae, sometimes called the gourd or marrow family, includes cucumbers, melons, pumpkins, squash, and all utilitarian, decorative, and edible gourds. Cucurbits are vines and usually monoecious, or producing male and female flowers on the same plant.

*Bitter Gourd (*Momordica charantia; *Cucurbitaceae)*

Bitter gourd is native to eastern Africa and India, specifically the northeast Himalayas. It is also known as bitter melon or balsam pear

TABLE 2.1. Five a Day: The Color Way—American Dietetic Association's Guide to Choice of Fruits and Vegetables for the Phytochemical Content and the Health Benefits They Offer

Color	**Phytochemical**	**Benefits**
Red	Lycopene and anthocyanins	Heart health, memory function, lower risk of some cancers, urinary tract health
Yellow	Vitamin C, carotenoids, and bioflavonoids	Lower risk of some cancers, heart health, vision health, healthy immune system
White	Allicin	Heart health, keep cholesterol levels healthy, lower risk of some cancers
Green	Lutein and indoles	Vision health, lower risk of some cancers, strong bones and teeth
Blue/purple	Anthocyanins and phenolics	Lower risk of some cancers, urinary tract health, memory function, healthy aging

Source: Produce for Better Health Foundation, http://www.5aday.com/html/consumers/colorway.php. Accessed June 20, 2007.

in English and *karela* in Hindi. All parts of the plant, including the fruit, taste bitter. It is commonly consumed at its unripe stage as a vegetable and used in a number of dishes such as soups, curries, and pickles. The fruits are also dried and preserved for use throughout the year. The plant grows in tropical areas of Asia, the Amazon, eastern Africa, and the Caribbean. It is cultivated throughout the world for eating as well as for medicine. The vine bears oblong fruits resembling a small cucumber with rough edges. The immature fruit is pale green, turning to orange yellow and then red as it ripens.

Bitter melon has been used in traditional and folk medicine in Brazil, China, Colombia, Cuba, Ghana, Haiti, India, Mexico, Malaysia, New Zealand, Nicaragua, Panama, Peru, and elsewhere to treat diabetes, as a carminative, and to treat colics (Satyawati et al. 1987; Yesilada et al. 1999). It is also used topically for treatment of wounds, internally and externally for the management of worms, parasites, and peptic ulcers; as an emmenagogue; and as an antiviral for measles and hepatitis. Bitter melon has many uses in traditional medicine (Giron et al. 1991; Lans and Brown 1998)—as an antidiabetic, abortifacient, anthelmintic, contraceptive, emmenagogue, antimalarial, laxative, galactogogue, and as a purgative and treatment for dysmenorrhea, eczema, gout, jaundice, abdominal pain, kidney stones, leprosy, leukorrhea, piles, pneumonia, skin rashes and psoriasis, rheumatism, fever, and scabies—that have turned the attention of a number of modern researchers toward this plant. This resulted in authentication of its use in diabetes and complications of diabetes (nephropathy, cataract, insulin resistance), as an antibacterial and antiviral agent (including HIV infection), as an anthelmintic and abortifacient, and in the treatment of peptic ulcers. Studies have shown its efficacy for treating cancer (lymphoid leukemia, lymphoma, carcinoma, melanoma, breast cancer, skin tumor, prostatic cancer, squamous carcinoma of tongue and larynx, human bladder carcinomas, and Hodgkin's disease; Basch et al. 2003; Ng et al. 1992; Raman and Lau 1996).

The immature fruits are rich in vitamin C, vitamin A, phosphorus, and iron. Biologically active chemicals present in bitter melon include glycosides, saponins, alkaloids, fixed oils, triterpenes, proteins, and steroids (Raman and Lau 1996). Specific phytochemicals identified in this vegetable include momorcharins, momordenol, momordicilin, momordicins, momordicinin, momordin, momordolol, charantin, charine, cryptoxanthin, cucurbitins, cucurbitacins, cucurbitanes, cycloartenols, diosgenin, eleostearic acids, erythrodiol, galacturonic acids, gentisic acid, goyaglycosides, goyasaponins, and multiflorenol (Husain et al. 1994; Murakami et al. 2001; Parkash et al. 2002; Yuan et al. 1999). The hypoglycemic properties of bitter melon are due to the charantins, peptides, and alkaloids (Raman and Lau 1996) concentrated in its fruit and hence the more pronounced hypoglycemic and antihyperglycemic activity of the fruit compared with other parts of the plant (Ali et al. 1993). The potential of bitter melon in prevention and delay in progression of diabetic complications (nephropathy, neuropathy, gastroapresis, cataract, and insulin resistance) have been demonstrated in experimental animals (Grover et al. 2001, 2002; Rathi et al. 2002; Vikrant et al. 2001). Other studies on humans show that the use of bitter melon improves glucose tolerance in diabetic subjects (Leatherdale et al. 1981); reduces postprandial serum glucose in as many as 86% of cases and fasting glucose in as many as 5% of cases (Ahmad et al. 1999); and increases glucose tolerance (Welihinda et al. 1986).

Extracts from the leaves and the entire plant demonstrated antimicrobial and antiprotozoal activity (Khan 1998) and immunostimulant effects in clinical and experimental settings (Cunnick et al. 1990). Other properties reported include anti-HIV activity of two bitter melon proteins, alpha- and beta-momorcharin (present in seeds, fruit, and leaves; Zheng et al. 1999) and antiherpes (Foa-Tomasi et al. 1982); antipoliovirus (Foa-Tomasi et al. 1982; Schreiber et al. 1999); anticancer (lymphoid leukemia, lymphoma, choriocarcinoma, melanoma, breast cancer, skin tumor, prostatic cancer, squamous carcinoma of tongue and larynx, human bladder carcinoma, and Hodgkin's disease; Basch et al. 2003; Battelli et al. 1996; Ganguly et al. 2000; Licastro et al. 1980; Ng et al. 1994); abortifacient and antifertility (Chan et al. 1984, 1985, 1986; Law et al. 1983); antiulcerogenic (Gurbuz et al. 2000; Matsuda et al. 1999; Yesilada et al. 1999); anthelmintic (Lal et al. 1976); antimalarial (Kohler et al. 2002; Munoz et al. 2000); antipsoriasis (Claflin et al. 1978; Takemoto et al. 1982; Vesely et al. 1977); analgesic and anti-inflammatory (Biswas et al. 1991; Choi et al. 2002); hypotensive and anti-prothrombin (Hayashi et al. 1994; Wang and Ng 2001); and hypocholesterolemic and antioxidant activities (Ahmed et al. 2001; Anila and Vijayalakshmi 2000; Dhar et al. 1999;

Jayasooriya et al. 2000; Noguchi et al. 2001; Platel et al. 1993; Singh et al. 1989). It also shows immunomodulatory activity that includes immunosuppressive and immunostimulatory effects (Cunnick et al. 1990; Leung et al. 1987; Spreafico et al. 1983; Zheng et al. 1999). Alpha-momorcharin has a combination of abortifacient, tumor suppressive, and anti-HIV properties (Ng et al. 1992).

Cucumber (**Cucumis sativus**; *Cucurbitaceae*)

Cucumber is widely cultivated in many countries for its crisp mild-tasting fruits consumed as a popular green salad. The plant is a trailing vine with creeping stems that cling to the ground or a support, alternate leaves that are palmate-lobate with a long petiole and rough leaf blades covered with rigid hairs. Tendrils spring from the base of the petioles, and these enable the plant to climb structures. Flowers have yellow petals. The fruits are cylindrical and long and often have a slightly lumpy skin. The oval seeds are embedded in a greenish white pulp. Durust et al. (1997) report the vitamin and mineral content of cucumbers (in mg/100 g fresh weight [FW]): ascorbic acid, 12; iron, 0.33; copper, 0.08; and zinc, 0.25. Constituents of cucumbers reduce the risk of renal cell carcinoma (Rashidkhani et al. 2005) and have antihyperglycemic (Roman-Ramos et al. 1995) and hypolipidemic (Sudheesh et al. 1999) properties, suggesting a nutraceutical potential for this already popular vegetable.

Ivy Gourd (**Coccinia grandis, C. indica, C. codifolia**; *Cucurbitaceae*)

Ivy gourd is native to parts of Africa and India. This vegetable is also known as scarlet gourd, or *parval* or *tinda* in Hindi and *Kovai* in Tamil. The plant is a perennial or semiperennial vine with simple, alternate, palmately lobed, heart-shaped leaves resembling that of an ivy plant. The flowers are solitary, large, white, and 3 to 4 cm across. The fruits are oval berries 25 to 60 mm long and 15 to 35 mm wide, green when unripe, turning a bright scarlet when ripe. The ripe fruits contain numerous small, pale, flattened seeds.

It is cultivated in several parts of Asia, Australia, the Caribbean, the Philippines, the United States, North Africa, the Middle East, Fiji, the West Indies, and tropical South America for its unripe fruits, which are used as a vegetable. The fruits are one of the most common vegetable crops in India and have been used since time immemorial as a medicine to control blood sugar. The leaves, shoots, and immature fruits are cooked as a vegetable and often pickled. The fresh fruits are also used as a vegetable in curries and stir-fries. They mix well with *dal sambar* (a southern Indian curry dish) and curries made with ground coconut. The mature fruits are also preserved as pickles or canned in brine.

In Ayurvedic medicine, the leaves and fruits are used in treating diabetes. The leaves can also be ground and mixed with turmeric (*Curcuma longa;* Zingiberaceae) and applied externally on joints for rheumatism. Other conditions for which the ivy gourd is used include the common cold, cough, swellings, headache, loss of taste, mouth sores, throat pain, congestion, and nausea and vomiting.

The renewed interest in screening plants having acclaimed health benefits, functional properties, and possible other beneficial uses includes the ivy gourd for exploiting the antidiabetic potency described in Ayurvedic texts. The leaves of this plant were evaluated for their nutritional value by Xu et al. (2003), who report the leaf protein quality to be superior to that of seven vegetables examined. The leaves had a protein index of 0.81 compared with that of egg and their ratio coefficient for amino acid was 71.946. The ivy gourd also contained the minerals potassium, phosphorus, iron, zinc, and selenium, and the vitamins thiamine, riboflavin, and niacin. An analysis of the leaves for antioxidant content by Chanwitheesuk et al. (2005) revealed ascorbic acid (17 mg%), xanthophylls (2.7 mg%), tannins (18 mg%), and total phenolics (75 mg%).

Ivy gourd has been studied for its hypoglycemic properties in experimental animals and human diabetic subjects. It "hold[s] promise as [an] . . . antidiabetic agent" (Platel and Srinivasan 1997). It has been proposed that the hypoglycemic influence of the fruit is due to its effect on enzymes in glucose metabolism (Kumar et al. 1993, 1997). Other properties described include the insulin-like activity of leaf extracts (Kamble et al. 1998) and depression of glucose synthesis and enhancement of glucose oxidation, which lower blood glucose (Hossain et al. 1992; Shibib et al. 1993) and offer hepatoprotection (Rao et al. 2005). Ivy gourd also has antimicrobial (Sudhakar and Rao 2005), anti-inflammatory and antinociceptive (Rao et al. 2004), and antimutagenic effects (Kusamran et al. 1998).

*Snake Gourd (***Trichosanthes anguina;** *Cucurbitaceae)*

Snake gourd is of Asian origin, specifically India. In Chinese medicine the pericarp of the fruit is used for relieving stagnancy of *qi* (vital energy) circulation in the chest and the seed is used as a moistening agent for treating dry constipation. The root is described as sweet, slightly bitter, and slightly cold, and is used for treating diabetes, skin swellings, pus formation, boils, and abscesses. The stalk and leaves of the plant are also used as a folk remedy for fever. Snake gourd is used to treat complaints such as lung heat, cough with phlegm, chest pains, lung ulcers, and constipation. The root is used to clear heat and stop thirst, boost the stomach, generate fluids, and treat yin deficiency.

The snake gourd is an annual climbing vine cultivated in all tropical parts of the world. The leaves are hairy and angular with five to seven lobes. The white male and female flowers are borne on the same vine. The fruits are 2 to 10 ft long, narrow and cylindrical with tapering ends, and have a waxy surface showing striations. To discourage their tendency to curl, immature fruits are tied to a stone. Young fruits are harvested for vegetable uses and are cooked as curried dishes. Its fruits are green when immature and turn orange to red when ripe. The seeds are embedded in soft pith within the fruit.

An analysis of snake gourd shows it to consist of much moisture and little protein, fat, fiber, or carbohydrate. It is rich in vitamins and minerals, especially calcium, phosphorus, and iron, and has a substantial amount of carotene but little thiamine, riboflavin, or niacin (Inoi et al. 2001; Khanum et al. 2000). Snake gourd seeds are rich in antioxidants and exhibit antibacterial activity (Harit and Rathee 1995; Yadav and Syeda 1994).

*Ridge Gourd (***Luffa acutangula;** *Cucurbitaceae)*

Ridge gourd, also called Chinese okra, is native to India and has been cultivated since ancient times. Ridge gourd is grown as a commercial vegetable crop in a number of countries, including India, China, El Salvador, North and South Korea, Taiwan, Guatemala, Colombia, and Venezuela. It is also grown in home gardens throughout the United States, Asia, and Latin America. The immature fruits are cooked as a plain vegetable or cooked with dal or coconut milk in curried dishes. The hard and crusty skin is removed and the fruit sliced into a number of savory and sweet dishes, chutneys, and pickles. It is also used in Ayurvedic medicine as an emetic, to treat skin disease, and as an expectorant. In Suriname traditional medicine a tea made out of the leaves is used as a diuretic.

Ridge gourd is a climbing vine that is trained on trellises in gardens. The stem is five-angled, hairy, and has three long, curled tendrils at each node. The leaves are green with a rough surface, simple, and broadly ovate, with a heart-shaped base tapering at the tip. The male and female yellow flowers are borne on the same plant (monoecious). The male flowers are large and occur in clusters (racemes), while the female flowers are smaller and borne as single solitary flowers with a tiny ovary attached at the base. The fruits are long, ellipsoid, club shaped, and have raised ridges. The fruits have numerous black seeds embedded in the pith.

The health benefits and functional properties of ridge gourd have been extensively studied by several researchers. Ridge gourd has been identified as an excellent source of the antioxidant alpha-tocopherol (Ching and Mohamed 2001); a number of flavonoids, including myricetin, quercetin, kaempferol, luteolin, and apigenin (Miean and Mohamed 2001); and dietary fiber (Khanum et al. 2000). Several proteins extracted from ridge gourd were ribosome inactivating and exhibited abortifacient, antitumor, and immunomodulatory activities as well as anti-HIV activity (Ng et al. 1992).

*Ash Gourd (***Benincasa hispida;** *Cucurbitaceae)*

Ash gourd is possibly a native of the Indo-Chinese–Indonesian region. The fruits, tender leaves, and twigs are consumed as leafy vegetables. Green fruits are used in making curry with fish or other vegetables or deep-fried. In West Bengal and Bangladesh, the mature fruits are mashed and mixed with black gram, a legume, to make *bori* and used as an ingredient in making an indigenous sweet called *morabba*. In Korean medicine it is described as efficacious against diabetes, dropsy, liver diseases, and leukorrhea and good for detoxication of minerals, lowering fever, and strengthening the function of the bladder and small and large intestines (Choi et al. 2001).

The plant is a climbing vine, with hairy stems and soft, pubescent, heart-shaped, five- to seven-lobed leaves. Flowers are large and yellow.

The fruits are 20 to 50 cm long and 10 to 25 cm across; ovoid-oblong, ellipsoid, or globose; and covered by a white ash like coating. The fruits are heavy, weighing up to 15 kg, and the oval seeds are embedded in the white pulp. Researchers have reported that seed extracts have an expectorant effect (Kim and Shin 1999), prevent gastric ulcer (Grover et al. 2000, 2001), inhibit histamine secretion (Yoshizumi et al. 1998), and show antitumor effects by enhancing immunoreactions and antiangiogenic properties (Lee et al. 2005).

BRASSICAS

The Brassicaceae is a large family comprising approximately 3,000 species that have been described and grouped into 350 to 380 genera. All members of the Brassicaceae (formerly the Cruciferae family) have flowers that are easily distinguishable: four petals and four sepals, six stamens (four long stamens and two short stamens), and two rows of seeds on a pistil partitioned lengthwise into two divisions, giving the appearance of a crucifix. Members of the *Brassica* genus include a wide range of horticultural crops (e.g., cabbage, mustard, pak choy, daikon, and broccoli), all of which contain the phytochemical compounds known as glucosinolates. Certain glucosinolates and their breakdown products, isothiocyanates, are linked to a reduction in the prevalence of certain types of cancers (e.g., gluconasturtin is linked to the reduction of cancer caused by tobacco-specific carcinogens). The antioxidant properties of cruciferous vegetables have been widely reported (Lampe et al. 2000). The glucosinolates and isothiocyanates in brassica vegetables can induce phase-2 enzymes that can inactivate carcinogenic enzymes (Potter and Steinmetz 1996; Lee and Lee 2006). Some of the glucosinolates that have recently received much publicity for anticancer action are glucoraphanin, found in broccoli, and phenethyl isothiocyanate, found in watercress. A number of *Brassica* species also contain high levels of antioxidant vitamins (alpha- and beta-carotene, alpha- and gamma-tocopherol, and ascorbic acid; Kurilich et al. 1999).

*Mustard (*Brassica juncea; *Brassicaceae)*

Mustard is one of the most widely used spices and condiments. It has been known to humanity since at least 5 BC, when it was reported as being used by the Africans, Chinese, Greeks, and Indians. Mustard is an essential spice used as oil, powder, whole seeds, or ground paste. Indians use the seeds as an everyday spice, and no dish is complete without a seasoning of mustard seeds spattered in heated oil. Mustard seeds contain a volatile oil that yields a pungent aroma and a strong bite.

In 2005 global mustard seed production was 621,023 Mt and the area under cultivation was 748,778 ha (FAOSTAT 2006). It is cultivated throughout Asia, in Bangladesh, India, Nepal, Pakistan, China, and Japan, and in central Africa as a salad crop for its edible leaves, roots, stems, and flowers and as a cash crop for oil extracted from its seeds. In India the seed is the major economic product. Extraction of oil from the seed is by rotary mill, expeller, and hydraulic processes.

It is an erect, perennial, herbaceous plant that is cultivated as an annual or a biennial. Mustard plants grow 0.9 to 2 m (3-6.5 ft) tall. Leaves appear clustered at the base of a short swollen stem. The leaves are 0.3 to 0.5 m (1-1.5 ft) long, sessile or with short petioles, elongated and oval, crinkled with one or two leaflets on each side. The bright yellow flowers are borne on flower stalks 3 ft (0.9 m) tall. The sickle-shaped green fruit pods are 2.5 to 5 cm long and contain about six to nine seeds.

In folk medicine, mustard has been prescribed for scorpion stings and snake bites, epilepsy, toothache, bruises, stiff neck, rheumatism, colic, and respiratory troubles. It is a strong emetic and rubefacient (an irritant) that draws the blood to the skin's surface to warm and comfort stiff muscles. Other medicinal uses of mustard are as an anodyne, purgative, diuretic, and stimulant and as a remedy for arthritis, footache, lumbago, and rheumatism (Duke and Wain 1981). In Chinese traditional medicine, seeds are used for tumors. Ingestion may impart a body odor repellent to mosquitoes, and leaves are applied to the forehead to relieve headache. In Korea the seeds are used for abscesses, colds, lumbago, rheumatism, and stomach disorders. The Chinese prescribe the leaves for bladder inflammation or hemorrhage. Mustard oil is also used for skin eruptions and ulcers.

Several kinds of mustard are manufactured commercially using mustard seeds in different forms and combinations: American ballpark-style mustard made from white seeds and blended with sugar and vinegar and colored with turmeric; Bordeaux mustard made from unhusked black seeds blended with unfermented wine; Dijon mustard made from husked black seeds blended with wine, salt, and spices; English mustard made from white seeds and with turmeric added for color; German mustard made as a smooth blend of vinegar and black mustard, varying in strength (Weisswurstsenf is a coarse-grained, pale, mild mustard); and Meaux mustard made from partly crushed, partly ground black seeds mixed with vinegar.

In Europe and the United States, mustard greens are eaten raw in salads or cooked as a potherb. Mustard greens have a peppery flavor that is enhanced in European cuisine with sweet, mild vinegar. It is common to mix mustard greens with another peppery green such as watercress or with milder greens such as collards and kale. In Asia leaves and stems are used in stir-fry and added to soups and stews. Leaves and stems are also pickled. The pungent seeds are used to season meats and other dishes. Mustard oil is used as a cooking oil, hair oil, pickling oil, lubricant, and cheaper substitute for olive oil. Mustard seeds are used primarily in pickles, relishes, and condiments where the appearance of whole seeds is an attraction. Ground mustard is a spice and condiment used to flavor food. Mustard oil is widely used in India as a cooking oil. Powdered mustard is used as an emulsifier in mayonnaise and salad dressings and for flavoring barbecue sauces, baked beans, many meat dishes, and deviled eggs.

The unique flavor of mustard and other members of the *Brassica* genus is due to the glucosinolates that release characteristic isothiocyanates when crushed. The isothiocyanates in mustard and other crucifers have been identified as potential chemopreventives of cancers of several organs and sites (Fahey et al. 2002; Khan et al. 1997; Terry et al. 2001). Mustard leaves and seeds are good sources of minerals (Elless et al. 2000), and animal models have shown them to possess hypoglycemic (Khan et al. 1995), hypolipidemic, and antioxidant (Khan et al. 1996a,b,c, 1997) functional properties.

*Radish (*Raphanus sativus; *Brassicaceae)*

Radish refers to the globular, red types commonly grown in the United States as well as the long white radish (daikon) indigenous to Asia. Daikon is the second largest vegetable crop in Korea because of its use in making kimchi. The pungent white radish varieties are cultivated as a major vegetable in Asia, while the other mild varieties are cultivated in Europe, America, and other parts of the world.

Different parts of the radish, including roots, seeds, and leaves, are used for medicinal purposes (Nadkarni 1976). Radish leaves are used in folk medicine as a laxative, stimulant, digestive aid, appetizer, and treatment for other stomach disorders (Kapoor 1990; Nadkarni 1976), pointing to the presence of gut-stimulatory constituents. It is processed into a range of fresh, dried, salted, and pickled products and constitutes about 60% of the processed vegetable market in Japan, where it is a part of the daily diet.

Radish roots are used mainly as a cooked vegetable and in salads. Young leaves, pods, roots, flowers, and seeds are all edible raw or cooked. In India the roots are finely grated and used to flavor homemade bread and in making deep-fried dishes and dips. The tender roots are preserved by salting and pickling or in sauerkraut. In some countries the radish roots are sliced and dried for use throughout the year. The leaves are considered good sources of calcium, phosphorus, iron, sodium, and vitamins. Radish roots are often added to fresh salads, cooked in stews, or grated as a garnish for dipping sauces or a relish.

In Asian traditional medicine, it is believed that consuming radish stimulates digestion and has a tonic and laxative effect. It is often prescribed to treat kidney stones, chest and nasal-passage congestion, and digestive problems. Other medicinal properties attributed to radish in various cultures include anthelmintic, antibacterial, antifungal, antiscorbutic, antispasmodic, astringent, cancer controlling and preventing, carminative, diuretic, expectorant, laxative, and stomachic, and it is used as a poultice. The whole plant is used to treat intestinal parasites; leaves, seeds, and roots are used to treat asthma and other chest complaints; juice of the fresh leaves is used as a diuretic and laxative; seeds are used as a carminative, diuretic, expectorant, laxative, and stomachic; and the leaves and roots are used to treat diarrhea and bronchitis. The root is crushed and used as a poultice for burns, bruises, and smelly feet.

Radish is a biennial that is grown as an annual. The plant grows 0.3 to 0.6 m (1-2 ft) long. Green leaves rise from the underground roots as clusters and form an aboveground rosette. The leaves are long, pubescent, and rough to the touch. The stem is fleshy and very short (0.2-0.6 cm). The roots are white or red, cylindrical, and fleshy with tiny rootlets arising from the surface. The roots are tender, cylindrical, and elongated and can weigh 1 to 40 lb depending on the cultivated variety. Radish is also grown as a companion plant in lettuce, pea, tomato, and cucumber crops. Flowers are white or lilac with yellow. The podlike fruit is constricted and bears numerous seeds.

A review of the large body of epidemiological data, including both cohort and case-control studies, indicates that consumption of cruciferous vegetables including radish has preventive potential for certain types of cancers. Table 2.2 provides a chemical analysis of the flavonoid content of radish leaves. Extract of radish is reported to be effective in treating obesity and diabetes (Kawabata and Kasai 2000) and as an antimicrobial agent (Bari et al. 1999). Raw radish sprouts have higher vitamin, mineral, and fiber content than cooked (Zielinski et al. 2005). Radish consumption also enhances excretion of calcium oxalates in the urine (Kumar 2004), adding to its health value.

*Chinese Cabbage (*B. chinensis; *Brassicaceae)*

Chinese cabbage is a nonhead cabbage, also called pak choi, bok choy, or Chinese white cabbage, of Chinese origin. It is a leafy vegetable widely grown in south China. Dark green leaves are borne on white spoon-shaped upright stems. The slight mustardy flavor of Chinese cabbage makes it a delightful addition to stir-fry, noodle and meat dishes, and soups, salads, and pickles. It is also dried or preserved as pickles. Chinese cabbage has a light, sweet flavor and crisp texture and contains minerals and vitamins, such as B_2, and antioxidant carotenoids (Wills and Rangga 1996). Regular consumption of Chinese cabbage and other brassicas is highly recommended for reducing the risk of colon cancer (Tan et al. 1999). Chinese cabbage could be an important source of dietary flavonols, in particular, kaempferol and isorhamnetin (Rochfort et al. 2006). Ching and Mohamad (2001) reported the alpha-tocopherol content (per 100 g dry weight [DW]) is 1.75 mg for nonhead cabbage and 0.049 mg for Chinese head cabbage (*B. pekinensis*).

ALLIUMS

Allium is the largest and the most important representative genus of the Alliaceae family, comprising 450 species, and is widely distributed

TABLE 2.2. Carotenoid Composition of Green Leafy Vegetables

Leafy Vegetables	Neoxanthin	Violaxanthin	Lutein	Zeaxanthin	Total Xanthophylls[a]	β-Carotene	Total Carotenoids[b]
B. rubra	12.01	6.947	67.94	2.25	89.14	32.42	121.56
M. oleifera	9.60	18.300	50.40	4.13	82.43	22.89	104.56
T. foenum-graecum	3.32	33.260	59.60	0.95	97.13	12.13	109.26
R. sativus	3.30	5.400	22.30	0.75	31.75	11.20	42.95

Source: Lakshminarayana et al. 2005.

Note: Values are milligrams per 100 g dry weight.

[a]Total xanthophylls = neoxanthin + violaxanthin + lutein + zeaxanthin.

[b]Total carotenoids = total xanthophylls + hydrocarbon carotenoid.

in the Northern Hemisphere. Besides the well-known garlic and onion, several other species are widely grown for culinary use, such as leek (*A. porrum*), scallion (*A. fistulosum*), shallot (*A. ascalonicum*), wild garlic (*A. ursinum*), elephant garlic (*A. ampeloprasum* var. *ampeloprasum*), chive (*A. schoenoprasum*), and Chinese chive (*A. tuberosum*). Many recent studies have suggested that certain species may prevent a number of chronic conditions and degenerative diseases such as carcinogenesis, atherosclerosis, and liver necrosis and have antifungal properties (Ghahfarokhi et al. 2006). Extensive research has focused on the beneficial and medicinal properties of several alliums, including garlic and onion. In particular, their use as an antioxidant in the treatment and prevention of cardiovascular disease and cancer is an area of considerable investigation and interest in current research (Ali et al. 2000; Augusti 1996; Beatty et al. 2000; Bordia et al. 1996; Chen et al. 2000; Fukushima et al. 1997). Alliums in general are recognized as rich sources of several phytonutrients contributing to the benefits of a Mediterranean diet and are also used in the treatment and prevention of a number of diseases, including cancer, coronary heart disease, obesity, hypercholesterolemia, diabetes, hypertension, cataract, and gastrointestinal-tract health. The sulfur compounds and phenolics in alliums have been reported to have antibacterial, hypoglycemic, hypocholesterolemic, hypotensive, anticarcinogenic, antimycotic, antiplatelet, antiatherosclerotic, antithrombotic, and antiasthmatic properties.

Garlic and onion were reported as hypoglycemic agents as long as 75 years ago (Janot and Laurin 1930; Laurin 1931). Perusal of the ancient literature finds evidence they were used in treating diabetes. Many claims for the healing and prophylactic properties of alliums are made on the basis of their continued use in India's indigenous systems of medicine.

Onion (**A. cepa**; *Alliaceae*)

Onion is native to Asia, Afghanistan, and Iran, and has been cultivated for more than 5,000 years to use as food and medicine. The name *onion* is derived from "union" and refers to the union of concentric layers of fleshy leaves that form the onion bulb. *Allium cepa* comprises a number of cultivated varieties of onions, bulb onions, shallots, pickling onions, tree onions, and scallions. The bulb and the green leaves of the plant are eaten in most world cultures as a vegetable and culinary herb. In many Indian dishes onion forms the base of curries and is a flavoring agent. However, because of its strong flavor, many Hindus despise onions and believe that it is not fit for sages and gods. Even today, traditional Hindus do not use onions in their diet because of the attributed *rajasic* property that would promote feelings of sensual pleasures.

In 2005 global production of dry onions was 57,909,848 Mt and the area of harvest was 3,180,629 ha, and of onions and shallots, 4,670,677 Mt and 231,400 ha, respectively (FAOSTAT 2006). Onion is cultivated in many parts of the world because of its status as a cash crop and acceptance as a daily vegetable. Main production countries are China, India, Pakistan, Iran, the United States, Turkey, and Japan. It is a hardy perennial or biennial but grown as an annual when cultivated for bulbs. Onion plants grow best under bright sunlight and in rich, well-drained, loamy soils with adequate irrigation.

Onion bulbs and leaves, eaten raw or cooked, are a versatile food as a vegetable and herb. It is sliced into salads or sandwiches and baked, fried, or boiled as a vegetable in its own right. It is also a commonly used flavoring in soups, stews, and many other cooked dishes, and it is pickled alone or with other vegetables. Onion flowers are used as a garnish on salads. The seeds have a delicious onion flavor and are often sprouted and eaten in some cultures.

Onion is used medicinally as an anthelmintic, anti-inflammatory, antiseptic, antispasmodic, carminative, diuretic, expectorant, hypoglycemic, hypotensive, and stomachic, and as a homeopathic and general tonic (Grieve 1982; Chevallier 1996). It is also useful in preventing oral infection and tooth decay. Baked onions are sometimes used as a poultice to treat skin infections (Chevallier 1996), and fresh onion juice is used on the skin to repel insects and as a treatment for bee and wasp stings, bites, grazes, or fungal infections. Drops of coconut or sesame oil in which onion pieces are boiled are used to treat earache. Onion bulbs from some red cultivars are used in homeopathic remedies to treat symptoms of the common cold and runny eyes and nose (Castro 1990). Onion juice rubbed on the skin is said to promote the growth of hair and to be a remedy for baldness (Chevallier 1996).

The onion plant is 0.3 to 0.6 m (1-2 ft) tall. The edible bulb is a modified shoot that forms underground, often very close to the surface. The leaves are hollow, rounded, and approximately 45 cm long. The swollen leaf base protects the bulb, forming thick scales around the fleshy bulb. The scales are attached to a basal stem from which the shallow root system rises. The root system is fibrous and shallow, spreading from just beneath the soil to 30 to 45 cm deep. The bulb comprises concentric, enlarged, fleshy leaf bases. The outer leaf base dries and becomes scaly and the inner leaf bases increase in thickness as the bulb develops. The stem of the onion is made up of hollow green leaves composed of a blade and a sheath. The sheath develops to completely encircle the growing point and forms a hollow tube enclosing the younger leaves and apex of the shoot. As more new leaves are initiated and grow, the older leaf bases get pushed farther away from the apex. The flowers (cymes) are borne on long stalks.

The medicinal properties and reported applications of onion include antimicrobial, antithrombotic, antitumor, hypolipidemic, antiarthritic, and hypoglycemic agents (Kendler 1987; Kumari et al. 1995; Roman-Ramos et al. 1995). Onion aids formation of scar tissue, thus speeding up the healing process, and is effective as a cosmetic to remove freckles (Chevallier 1996).

Onion bulbs and leaves are important sources of dietary flavonoids, containing a range of quercetin, isorhamnetin, and kaempferol conjugates (Leighton et al. 1992; Tsushida and Suzuki 1995). The highest total flavonoid content was in onion leaves (1,497.5 mg/kg quercetin, 391.0 mg/kg luteolin, and 832.0 mg/kg kaempferol).

Onion has many sulfur-containing active principles, mainly in the form of cysteine derivatives, which on extraction decompose into thiosulfates and polysulfides by the action of the enzyme allinase. Onion extracts have antibiotic, hypocholesterolemic, fibrinolytic, and various other biological actions (Augusti 1996). Onion has also been reported to be hypotensive, hypolipidemic, and antidiabetic (Goldman et al. 1996; Helen et al. 1999; Mayer et al. 2001) and effective in controlling oral pathogenic bacteria (Kim 1997).

The earliest experimental evidence on the hypoglycemic effect of onion came in 1961 (Brahmachari and Augusti 1961, 1962). Antidiabetic and antiperoxide effects of *S*-allyl cysteine sulfoxide were reported by Augusti and Sheela (1996). Other studies provide evidence that eating onions had significant hypolipidemic, antioxidant, and hypoglycemic influence and caused amelioration of diabetic nephropathy in diabetic rat models (Suresh and Srinivasan 1997, 1999).

*Garlic (***A. sativum***; Alliaceae)*

Garlic is a food and a medicinal herb indigenous to India and east Asian countries. It has been used since the earliest Chinese dynasties and the time of the Egyptian pharaohs. It is also mentioned in the Bible as a food the Hebrews enjoyed during their sojourn in Egypt and in the writings of Pliny the Elder, the first-century Roman naturalist (AD 23-79). Garlic is used in all parts of the world, not only as a spice and food but also as a folk remedy for ailments such as heart disease, cancer, diabetes, infections, parasites, respiratory ailments, and digestive problems and to delay blood clotting.

Garlic is well known for its strong characteristic odor, which is distinctly different in fresh and cooked states. While garlic is the spice of life and is revered around the world for its potent health benefits, conservative Hindus despise it for its odor as well as its *rajasic* properties, which enhance sensual feelings, and deem it unfit for gods and sages.

In 2005 global garlic production was 14,556,917 Mt and the area of harvest was 1,129,750 ha (FAOSTAT 2006). Garlic is cultivated throughout the world, including Austria, Britain, Bulgaria, Canada, China, Cuba, Egypt, France, Haiti, India, Iraq, Malaysia, Mexico, Pakistan, Panama, Paraguay, Puerto Rico, Russia, South Africa, Spain, Trinidad, Turkey, and the United States. Garlic is used as a spice and vegetable in cooking and to flavor butter, snacks, vinegar, oil, and salt. Asian cuisine makes use of fresh garlic and pastes of garlic; combined with ginger, chilies, or lemongrass it is added to soups or stews in Vietnam, India, Cambodia, Thailand, and Indonesia. Fresh garlic cloves are often pickled and used in preserves. Fresh garlic leaves are also edible and chopped into salads.

The health benefits of garlic listed in world literature are many. Garlic was recommended as early as first century AD by Pliny the Elder as an antidote for poisonous bites of shrews and snakes, in the treatment of asthma, as a cough suppressant, to expel intestinal parasites, and as an aphrodisiac when mixed with freshly crushed coriander seeds and wine. Garlic is used medicinally in many traditional folk medicines to treat colds, coughs, high blood pressure, digestive disorders, fungal infections, bacterial infections, cancer, high blood sugar, high cholesterol levels, roundworms, and hookworms (Grieve 1982; Mabey 1988; Murray 1995; Ody 1993). According to Ayurveda, garlic has a heating property and is a diuretic, diaphoretic (promoting sweating), expectorant, and carminative. It also has immune-boosting, anticoagulant, and anthelmintic effects. Fresh cloves are rubbed on acne, diluted garlic juice is used to clean fungal infections, and it is taken internally for intestinal worms and indigestion. It is a common practice to make poultices by frying garlic and onions in olive oil or coconut oil to apply to the chest or back to relieve lung congestion or to put in ears to cure earaches and ear infections. It is prescribed in Chinese medicine and in homeopathy for a number of ailments, including tuberculosis, coughs, and digestive complaints such as diarrhea.

The garlic plant is a small herbaceous plant that grows 0.3 to 0.6 m (1-2 ft) tall. The leaves are green, slender, flat, and bladelike measuring 30 to 45 cm (1-1.5 ft) long. The stem is smooth and solid. The inflorescence is an umbel enclosed in a spathe. The edible underground bulbs are made of individual cloves held together in a rounded bulb. The bulbs are composed of several cloves, encased in white or pale pink scalelike skin.

Modern research on garlic has focused on its value as a nutraceutical and a functional food. Garlic has become recognized for its great value in the prevention of a number of chronic conditions related to heart health, including arteriosclerosis. The health benefits of garlic in human health are attributed to its sulfur-containing compounds—allicin, diallyl disulfide, diallyl trisulfide, alliin, and *S*-methyl-L-cysteine sulfoxide (Milner and Schaffer 1998).

More than 3,500 peer-reviewed scientific articles have been published on all aspects of garlic, including its chemistry, pharmacology, and clinical and cosmetic applications. Clinical trials show garlic extracts reduce blood viscosity, inhibit platelet aggregation, and increase fibrinolytic activity (Bordia et al. 1998). Garlic is also a strong antifungal agent against *Candida albicans* and an effective antibiotic against *Staphylococcus aureus, Streptococci* spp., *Escherichia coli, Proteus vulgaris, Salmonella enteritidis,* and *Mycobacteria* spp. in the digestive tract, aiding maintenance of intestinal good health.

Garlic also protects against stroke and myocardial infarction by preventing clot formation through inhibition of platelet aggregation and increased fibrinolysis (Murray 1995). The role of garlic in prevention and treatment of cancer has received increasing attention in the past decade. Numerous epidemiological studies show an inverse relationship between garlic consumption and the incidence of cancer. Experimental studies with animal models have shown garlic extracts to have potent chemopreventive and anticancer effects (Borek 2001; Dirsch et al. 1998; Dorant et al. 1993; Fleischauer et al. 2001; Fukushima et al. 1997; Ip et al. 1996; Lamm and Riggs 2000; Milner 1996).

Garlic has exhibited anti-inflammatory action in animal models and antioxidant activities, particularly related to reduced glutathione (GSH) enhancement. Garlic extracts are hypotensive (Pedraza-Chaverri et al. 1998) and capable of immune modulation (Lau et al. 1991; Lamm and Riggs 2001). Garlic indirectly affects atherosclerosis by reduction of hyperlipidemia, hypertension, and probably diabetes mellitus and prevents thrombus formation. In addition, in animal models garlic causes direct antiatherogenic and antiatherosclerotic effects at the artery wall. Garlic's direct effect on atherosclerosis is due to its capacity to reduce lipid content in arterial cells (Orekhov and Grunwald 1997; Phelps and Harris 1993; Pinto et al. 1997; Silagy and Neil 1994; Yeh and Liu 2001) and decrease cholesterol level in blood (Aouadi et al. 2000; Stevinson et al. 2000).

Some researchers have reported garlic to be antidiabetic, with its ability to lessen moderately elevated blood sugar levels (Steiner et al. 1996). Sheela et al. (1995) and Augusti and Sheela (1996) observed that *S*-allyl cysteine sulfoxide isolated from garlic caused a reversal of diabetic conditions such as glucose tolerance and weight loss.

Though garlic is among the most medicinally active substances consumed by people the world over, with documented antimicrobial, antithrombotic, and anticancer activities (Nagourney 1998; Milner 1996), some adverse effects have been reported, perhaps due to consumption in excess or allergic reactions in specific individuals. Adverse effects of garlic that have been reported after oral ingestion, inhalation, and topical exposure include smelly breath, body odor, flatulence, esophageal and abdominal pain, small intestinal obstruction, contact dermatitis, rhinitis, asthma, bleeding, myocardial infarction, urticaria, angioedema, anaphylaxis, skin blisters, and ulcero-necrotic lesions (Morbidoni et al. 2001).

GREEN LEAFY VEGETABLES

Leafy vegetables are easy to grow and inexpensive, rich sources of several nutrients. Worldwide they are consumed and used for their desirable chemical composition and medicinal properties. In tropical countries green leafy vegetables are important sources of vitamins and minerals and are included in the daily diet.

*Chekup Manis (*Sauropus androgynus; *Euphorbiaceae)*

Also called chekurmanis, *chekkurmenis,* sayor manis, cekor manis, and tropical asparagus, chekup manis has been a popular leafy green perennial vegetable in Malaysia for many years. It was introduced into India in the 1950s and has come to be known as the multivitamin plant. The leaves have been used for decades in West Java as an herbal medicine to increase breast milk (Hasanah et al. 1999).

The plant grows to about 3.5 m tall. The leaves are alternate, oblong, 2 to 8 cm long and 1 to 4 cm wide, glaucous (waxy coating) with entire margins. Flowers appear at axils and mainly form clusters. The fruit is a fleshy berry, ovoid or globose.

It has a pleasant taste, similar to fresh garden peas, and a slightly nutty flavor. The leaves and the top 15 cm of the stem tips are normally eaten raw in salads or steamed, or added to stir-fry, rice, egg dishes, soups, or casseroles (Fletcher 1998). The leaves retain their dark green color and firm texture on cooking and are served in restaurants as *sayor manis*. The flowers and small purplish fruits of the plant are also eaten. Chekup manis has a high level of provitamin A carotenoids, especially in freshly picked leaves, as well as high levels of vitamins B and C, protein, and minerals. Nutrient content rises as the leaves mature; 100 g of fresh leaves have 11.6 g carbohydrate, 6.8 g protein, 370.0 mg calcium, 28.0 mg iron, and 247.0 mg ascorbic acid (Hemalatha et al. 1999). Ramachandran et al. (1980b) measured 9,510 IU of vitamin A and 247 mg vitamin C. Ching and Mohamed (2001) report tocopherol at 426.8 mg/kg edible portion. Giri et al. (1984) report each 100 g to have beta-carotene, 4.62 to 6.00 mg; ascorbic acid,

238 to 257 mg; riboflavin, 0.26 to 0.33 mg; thiamine, 0.28 to 0.49 mg; niacin, 1.59 to 2.99 mg; phosphorus, 179 to 219 mg; calcium, 0.393 to 0.593 mg; magnesium, 0.362 to 0.616 mg; sodium, 0.098 to 0.209 mg; potassium, 0.975 to 1.308 mg; zinc, 0.199 to 0.254 mg; iron, 18.9 to 28.4 mg; copper, 0.235 to 0.270 mg; and manganese, 2.24 to 2.80 mg. It has antioxidant potential against breast tumors (Rahmat et al. 2004) and hypoglycemic activity (Sai and Srividya 2002).

*Vegetable Drumstick (***Moringa indica, M. oleifera;** *Moringaceae)*

Drumstick is native to India and possibly parts of Africa, Arabia, Southeast Asia, South America, and the Pacific and Caribbean islands. The drumstick tree is also called the ben oil tree, after the commercial oil extracted from the seeds, or the horseradish tree, because its root substitutes for horseradish. The drumstick's leaves, fruits, and flowers are edible and a common vegetable in India, the Philippines, Hawaii, and many parts of Africa and Asia. Drumstick tree leaves are eaten as greens in salads and in vegetable curries in Malaysia and India. They are applied as a poultice to sores, rubbed on the temples for headaches, and said to have purgative properties (Duke 1983a).

The drumstick tree is widely cultivated in tropical Africa, tropical South America, Sri Lanka, India, Mexico, Malaysia, and the Philippines. The drumstick leaves and pods are used as a vegetable, an aphrodisiac, food, and medicine in Asia (Ramachandran et al. 1980a). The flowers are either mixed and cooked with other foods or fried in batter. Tender leaves are used to flavor ghee and to enhance the shelf life of ghee. In rural areas in Sudan, ground drumstick seeds are suspended overnight in turbid household water to purify it (Jahn 1981; Jahn et al. 1986).

Almost all parts of the tree are utilized in Ayurvedic medicine and other folk remedies. According to Hartwell (1967-1971), the drumstick flowers, leaves, seeds, and roots are used for tumors. The roots are bitter, are used as a tonic to the body and lungs, an expectorant, a diuretic, and as a stimulant, and are used to treat paralytic afflictions, epilepsy, and hysteria. Juice prepared from the root is applied externally to heal skin irritation. The leaves have purgative properties and are applied as a poultice to sores, rubbed on the temples for headaches, and used for piles. The bark, leaves, and roots of the drumstick tree are used as a digestive aid.

The drumstick tree is a perennial that grows up to 6.1 m (20 ft) tall, with a straight trunk and corky, whitish bark. The leaves are pale green, compound, tripinnate, 30 to 60 cm (11.8-23.6 in) long, with many small leaflets. The lateral leaflets are elliptic, and the terminal one is obovate and slightly larger than the lateral ones. The fruit pods are pendulous, green turning greenish brown, and triangular and split lengthwise into three parts when dry. The pods are 30 to 120 cm (1 to 4 ft) long and 1.8 cm (0.7 in) wide, tapering at both ends. The pods contain about 10 to 20 seeds embedded in the fleshy pith. The seeds are dark brown, and the kernel is surrounded by a lightly wooded shell with three papery wings.

The fresh tender leaflets are stripped and used as a leafy vegetable and a potherb that is cooked alone or with lentils. The young pods, also known as drumsticks or *moringai* by Indian communities, are cooked as a vegetable. The pods are sold fresh or canned in many Asian grocery stores and markets. The green peas and surrounding white material can be removed from larger pods and cooked in many ways. The pods are considered good sources of essential amino acids. A compound found in the flowers and roots of the drumstick tree, pterygospermin, has powerful antibiotic and antifungal effects (Das et al. 1957).

Drumstick leaves are a good source of protein (Freiberger et al. 1998). The leaves, flowers, and pods are significant sources of vitamins A, B, and C, riboflavin, nicotinic acid, folic acid, pyridoxine, ascorbic acid, beta-carotene, calcium, iron, and alpha-tocopherol (Dahot 1988). Leaves of *M. oleifera* have been reported to contain flavonoid pigments such as kaempferol, rhamnetin, isoquercetin, and kaempferitrin (Nair and Subramanian 1962). Aqueous leaf extracts are reported to regulate thyroid hormone and can be used to treat hyperthyroidism (Tahiliani and Kar 2000). Leaf extracts are also used to treat ulcers (Pal et al. 1995). Drumstick leaves and pods reduce blood cholesterol (Ghasi et al. 2000) and have antitumor activity (Guevara et al. 1999). Drumstick leaves and pods are an important source of the glucosinolate precursors of the isothiocyanate group of chemopreventives (Daxenbichler et al. 1991) that can inhibit carcinogenesis. Drumstick is also being studied for its anti-inflammatory, antimicrobial, diuretic

(Caceres et al. 1991, 1992), antibiotic (Eilert et al. 1981), hypotensive (Faizi et al. 1998), antimicrobial (Spiliotis et al. 1998), and antioxidant properties (Siddhuraju and Becker 2003).

Resmi et al. (2005) analyzed the moringa: Vitamin A in leaves was 8,108 to 13,216 IU and in fruit 95 to 185 IU, vitamin C in leaves (per 100 g) was as high as 226 mg and in fruit 129 mg. Sánchez-Machado et al. (2006) found drumstick a good source of tocopherols: alpha-tocopherol (in µg/g DW) was 745 in mature leaves, 306 in flowers, and 96 in beans. Aslam et al. (2005) reported considerable variability in the nutrient quality of the leaves and pods of drumstick tree: Concentrations (in mg/kg) in leaves were 9,732 to 24,397 potassium, 1,839 to 2,097 calcium, 18,950 to 26,349 magnesium, and 1,292 to 1,837 sodium versus 98.2 to 109, 93.9 to 103.9, 1,635 to 2,721, and 1,032 to 2,105, respectively, in pods. The concentrations (in mg/kg) in leaves were 205 to 573 iron, 155.2 to 435.9 copper, 7.3 to 11.2 manganese, and 20.9 to 32.1 zinc versus 76.9 to 112.8, 40.2 to 72, 20.9 to 34.1, and 15.3 to 29, respectively, in pods. The concentration of phosphorus in leaves and pods was 1,180 to 1,450 and 1,860 to 2,125 mg/kg, respectively.

The drumstick also exhibited high phenolic content and antioxidant activity due to high concentrations of kaempferol (2,360 µg/g leaf DW) and quercetin (207 µg/g leaf DW) (Bajpai et al. 2005; Siddhuraju and Becker 2003). Drumstick's anticancer potential has been documented (Lotufo et al. 2005) as well as its nootropic property (Mohan et al. 2005) in Asian folk medicine. The antioxidant activity of leaves might be caused by the presence of high amounts of flavonoids, polyphenolics, and tocopherol content (see Table 2.2). *Moringa oleifera* leaves are described as a viable source of natural antioxidants and nutraceuticals (Siddiq et al. 2005). The leaves are used in rural India to treat joint diseases such as gout, arthritis, and rheumatism (Dwivedi et al. 2005). Leaf powder is recommended as a useful, simple, effective, easily available, and economically affordable supplement for the effective management of non-insulin-dependent diabetes mellitus (John and Chellappa 2005).

*Fenugreek (***Trigonella foenum-graecum***; Fabaceae)*

Fenugreek is native to southern Europe, the Mediterranean region, and western Asia. Fenugreek is widely cultivated as a spice, fodder crop, and medicine in many tropical and subtropical regions, including India, North Africa, and the United States. The leaves, both fresh and dried, are eaten as a leafy vegetable component in curries, cooked with *dal,* or mixed with wheat flour and made into *methi chapati,* a flat wheat bread in India consumed almost every day in some parts of India. Because uncooked fenugreek seeds have an unpleasant, bitter taste, even after roasting and grinding them to a powder, they are used in very small quantities. The seeds are also used extensively in chutneys, pickles, candy, baked goods, ice cream, chewing gum, and soft drinks.

In Ayurvedic and Unani systems of medicine, fenugreek treats epilepsy, paralysis, gout, dropsy, chronic cough, and piles. In Indian traditional medicine fenugreek has a cooling effect on the human body and has been used as an expectorant, a restorative tonic, and most effectively in diabetes management. Decoctions of fenugreek seeds are given to treat diarrhea and dysentery, for bronchitis or fevers, and to treat sore throat. A poultice of crushed seeds is used for swellings, tumors, sores, and skin irritations. Fenugreek powder is a natural shampoo and conditioner and a skin cosmetic. Traditionally it has been used to stimulate milk production in nursing mothers and cure menstrual cycle problems. Traditional Chinese herbalists prescribe fenugreek for kidney problems and conditions affecting the male reproductive tract.

Fenugreek is an annual herb that grows to 0.3 to 0.9 m (1-3 ft). The stem is erect with green trifoliate leaves on hairy petioles, with obovate leaflets. The flowers are small and white and have three corners. The dark yellow or mustard-colored fruits contain 10 to 20 small, flat, pungent, aromatic seeds in a pod. The seeds have a strong, distinct aroma and bitter taste. Dried fenugreek seeds are small, hard, angular, and yellowish brown.

This time-tested folk remedy for diabetes and a number of chronic disorders has been studied extensively by modern methods for its chemical composition and health benefits. Choudhary et al. (2001) describe fenugreek's activities as a modulator of several antioxidant enzymes in the liver, spleen, and erythrocytes. Antidiabetic and hypocholesterolemic properties of fenugreek seeds have been demonstrated by researchers in animals and humans (Bordia et al. 1997; Madar et al. 1988; Raghuram et al. 1994; Rao et al. 1996; Ravikumar and

Anuradha 1999; Sharma et al. 1990, 1991, 1996; Zia et al. 2001). Fenugreek seeds contain a mucilaginous fiber with high viscosity that provides fiber, increases beneficial bacteria in the human gut (Madar and Thorne 1987; Al-Khaldi et al. 1999), and is beneficial in preventing heart disease (Vajifdar et al. 2000). Some have found fenugreek seeds effective in preventing microbial growth and preserving meat (McCarthy et al. 2001a,b; Mansour and Khalil 2000), suggesting its potential use as a food preservative. Other medicinal properties demonstrated by fenugreek include antipyretic, anti-inflammatory (Ahmadiani et al. 2001), antioxidant, antitumor (Sur et al. 2001), and immunomodulating (Hafeez et al. 2003), the latter two being a relatively new source of chemopreventives against cancer. Fenugreek seeds are used as a condiment and a supplement to wheat and maize flour in bread making and as a constituent of India's daily diet. Fenugreek seeds contain tannic acid; fixed and volatile oils; a bitter extractive, diosgenin; the alkaloids trigonelline, trigocoumarin, trigomethyl coumarin; steroidal saponins such as gitogenin; and traces of trigogenin and vitamin A (Jayaweera 1981; Petit et al. 1995). Sharma et al. (1990) demonstrated in a human clinical trial that administration of fenugreek seed powder (50 g with each lunch and dinner) to insulin-dependent (type 1) diabetic patients for 10 days significantly reduced fasting blood sugar, improved glucose tolerance, caused a 54% reduction in glycosuria, and had a significant hypolipidemic effect.

Fenugreek leaves are consumed widely in India as a green leafy vegetable and are a rich source of calcium, iron, and beta-carotene and other vitamins (Sharma et al. 1996). A chemical analysis of fresh fenugreek leaves by Bajpai et al. (2005) found protein (4.7%), carbohydrates (2.7%), vitamin C (48.3 mg/100 g), and carotenoids (22.9 mg/100 g). Some of the therapeutic uses of fenugreek leaves are as a hypoglycemic, antiulcerogenic, hypocholesterolemic, and antihypertensive agent (Kirtikar and Basu 1935; Sharma et al. 1996; Tayyaba et al. 2001), antioxidant (Ravikumar and Anuradha 1999), and hypocholesterolemic (Stark and Madar 1993). See Table 2.2 for fenugreek's flavonoid content.

*Indian Spinach (***Basella rubra** *L.,* **B. alba** *L.; Basellaceae)*

Indian spinach, also called Ceylon spinach, Malabar spinach, Malabar nightshade, and vine spinach, is native to India and Africa.

Ceylon spinach is widely distributed in the tropics, particularly in India, Malaysia, the Philippines, China, Africa, the Caribbean, and tropical South America. It is a leafy vegetable grown in home gardens and well known as an excellent source of vitamins and minerals. The edible leaves and stems are mucilaginous, have a mild flavor, and are cooked as a potherb, often with *dal,* onions, and chilies; used in soups and stir-fries; or dipped in spicy batter and deep-fried. The fruit juice is sometimes used as a dye and a natural food color.

In Indian Ayurvedic and Siddha medicine the leaf and stem are useful as a mild laxative, diuretic, and demulcent and are proposed to have a cooling effect on the body. It is prescribed for excessive heat, for problems related to the uterus, and as an aphrodisiac to treat male sexual weakness. The leaves are used for oral care in treating apthae in some parts of India (Hebbar et al. 2004).

There are two common species of Ceylon spinach, one with red stems and leaves (*B. rubra*) and another with green stems and leaves (*B. alba*). It is a climbing perennial plant, mostly cultivated as an annual leafy vegetable. The fleshy leaves are ovate with short petioles. The leaves are alternately arranged on succulent stems. The pale yellow or green flowers borne on axillary spikes are very inconspicuous. The fruits are fleshy and purplish black. Ceylon spinach has been recognized as an "emerging" vegetable and widely studied for its nutritive value as a source of vitamins and minerals (Bhattacharjee et al. 1994; Rajyalakshmi et al. 2001), dietary fiber (Punna and Rao 2004), and lutein (68 mg/100 g DW) and beta-carotene (32.4 mg/100 g DW; Rangaswamy et al. 2005). Corlett et al. (2002) identified *B. alba* as an "exotic culinary-medicinal species" in the diet of Hmong residing in California and found it had high mineral profiles: it contains (in µg/g DW) 1,030 calcium, 1,142 magnesium, 3.1 iron, 2.79 manganese, and 0.99 zinc. A chemical analysis of Indian spinach leaves is presented in Table 2.2.

Some researchers have reported *Basella* spp. extracts as showing antioxidant activities (Chu et al. 2000; Yen et al. 2001). Leaf extracts of Indian spinach increased the serum level of testosterone, increased sperm count, and demonstrated androgenic potential in animal models (Moundipa et al. 1999, 2005, 2006), providing evidence for its traditional medicinal use in treating male sexual weakness. Daily consumption of Indian spinach leaves also increased vitamin A levels in populations at risk of vitamin A deficiency (Hasekell et al. 2004).

The seeds contain fatty acids, including palmitic, oleic, linoleic, linolenic, and arachidonic acids (Banerjee et al. 1992). A red pigment extracted from the ripe fruit dissolves easily in water and turns acid and neutral media bright purple; it is recommended as a safe, nonpoisonous food color and additive (Hu et al. 1989).

Ponnam Kanni (**Alternanthera sessilis**; *Amaranthaceae*)

Ponnam kanni is native to Southeast Asia (China, Malaysia, and probably the Philippines; Smith 1981). It is a perennial herb whose stems are prostrate, rooting at the nodes, and whose leaves are obovate to broadly elliptic, occasionally linear-lanceolate. It flowers in sessile spikes, with bracts and bracteoles that are shiny white. It is a common roadside weed in many Asian countries and the tropics. Leaves contain 5.7 mg/100 g beta-carotene (Bhaskarachary et al. 1995) and 22 mg/g DW of alpha-tocopherol (Sridhar and Lakshminarayana 1993). It is hypoglycemic (Sreedevi and Chaturvedi 1993) and antibacterial (Kumaresan et al. 2001; Sahu and Chakrabarty 1994) and used in the treatment of rickets (Agarwal and Arya 2005).

Purslane (**Portulaca oleracea**; *Portulacaceae*)

Purslane is an herbaceous annual with fleshy succulent leaves. It is a potherb native to the Indian Subcontinent. Purslane is reported in the literature as an invasive weed, a common potherb, and an unusual medicinal herb. Purslane is distributed worldwide, including the Mediterranean region, central Europe, North America, Africa, Asia, and Latin America. Purslane has been ranked as the eighth most common plant in the world (Coquillat 1951) and one of the three most frequently reported weeds across the world (Holm et al. 1977). Purslane is a commercially cultivated vegetable in many parts of the world, including Asia and the Middle East (Kays and Dias 1995), and is used as a vegetable in southern Europe, Mediterranean countries, and is Asia. It is palatable, has a mild flavor, and has a mucilaginous quality. Tender stems and leaves can be eaten raw, cooked, or pickled. The leaves of purslane can be frozen or dried and stored in jars for year-round usage.

The medicinal and health benefits of purslane are well documented. It is used to treat burns and trauma; headaches; stomach, intestinal, and liver ailments; cough; shortness of breath; and arthritis. It is used as a purgative, cardiac tonic, emollient, muscle relaxant, anthelmintic, anti-inflammatory, antiscorbutic, cathartic, and diuretic (Simopoulos et al. 1992).

Purslane is an annual herb with succulent stems tinted pink. The leaves are alternate or opposite, fleshy, and obovate. The wild type has smaller leaves and is of prostrate growth habit, while the cultivated types are erect and produce larger leaves. The plant produces small, sessile, yellow flowers that contain five bilobed petals. The small seed capsules produce abundant black seeds.

Purslane has attracted recent attention from biomedical, nutritional, and agricultural researchers (Adams 1992; Palaniswamy 1998; Simopoulos 1999) as a potential addition to the Western diet, because of its high concentration of essential fatty acids—linoleic and alpha-linolenic acids. In the United States it is grown as a specialty vegetable and is available as mesclun salad in health food stores.

Approximately one serving of fresh purslane leaves (100 g, or one 8-oz cup) can supply 300 to 400 mg alpha-linolenic acid, 12.2 mg alpha-tocopherol, 26.6 mg ascorbic acid, 1.9 mg beta-carotene, and 14.8 mg glutathione (Simopoulos et al. 1992). Purslane is also an excellent source of essential amino acids (Miller et al. 1984) and noradrenaline (Feng et al. 1961). Because of its high nutritive and antioxidant properties, purslane is described as a "power food of the future" (Levey 1993). Alpha-linolenic acid concentration in purslane is the highest of any green leafy vegetable examined to date and is reported as being higher than most commonly consumed fish (Simopoulos et al. 1992; Simopoulos and Salem 1986). Purslane appears to be the only higher plant reported to contain beneficial omega-3 fatty acids—eicosapentaenoic acid, docosahexaenoic acid, and docosapentaenoic acid (Omara-Alwala et al. 1991; Simopoulos and Salem 1986), which are precursors to a specific group of prostaglandins.

Omega-3 fatty acids have antiarrhythmic, anti-inflammatory, antithrombotic, hypolipidemic, and vasodilatory properties beneficial in patients with hypertension, type 2 diabetes, and coronary heart disease; it is used for patients with renal disease, rheumatoid arthritis, and ulcerative colitis (Simopoulos 1999). Other prospective uses of cultivated purslane are as an animal feed (Bosworth et al. 1980) and in the food-processing industry as a food extender and thickening agent (Wenzel 1990).

West Indian Pea (**Sesbania grandiflora**; *Fabaceae*)

West Indian pea is also known as hummingbird tree, sesban tree, corkwood tree, or *agathi* in different parts of Asia. It is native to India, Malaysia, Indonesia, and the Philippines. It is an everyday green vegetable much sought after in many parts of Asia for general well being, disease prevention, and health promotion. The West Indian pea is cultivated in the tropics for its valuable edible leaves used as a vegetable and as fodder for cattle. It is grown in fields and along roadsides and is cultivated in backyard vegetable gardens in Mauritius, India, Malaysia, the Philippines, Mexico, the West Indies, and many countries of Central and South America.

West Indian pea tree leaves are commonly used in Indian cuisine. The tender leaves, green fruit, and flowers are eaten alone as a vegetable or mixed into curries or salads. The young leaves are stripped from the stalks and lightly boiled or steamed with onions and spices or cooked as a vegetable in curries and salad. The young pods are also eaten as a vegetable. Flowers are fried in butter or deep-fried as a dip. Tender leaves, fruits, and flowers are also used as fodder.

In Asian traditional medicine agathi is believed to be a diuretic, emetic, emmenagogue, antipyretic, laxative, anthelmintic, and tonic. *Agathi* is used as a folk remedy for bruises (Duke 1983b), catarrh, dysentery, eye sores, fevers, headaches, smallpox, mouth sores, and sore throat (Duke and Wain 1981). It is also believed to improve vision and prevent eye diseases. Leaves are used as a poultice to treat bruises and swellings. Ayurvedic as well as Unani medical practitioners routinely prescribe agathi fruits and flowers for anemia, bronchitis, fever, and mouth sores; use the roots for inflammation; and use the bark as an astringent. Leaves are believed to have alexiteric and anthelmintic properties, and they are used for epilepsy, gout, itch, leprosy, and several disorders related to the eye and vision. The seed oil, which contains saturated and unsaturated fatty acids, is used in some Ayurvedic medicinal preparations. In other parts of Asia the flower juice is squeezed into the eye to correct dimmed vision; bark infusions are used for smallpox, diarrhea, and dysentery; and the flowers are used as a laxative. In Malaysia the crushed leaves are applied to sprains and a gargle of the leaf juice is used to cleanse the mouth and throat (Duke and Wain 1981). Bark extracts are used as a laxative and emetic, used to treat ulcers of the mouth and digestive tract, and applied externally to scabies. Leaves are chewed to disinfect the mouth and throat. The leaves and flowers contain vitamins A, B, and C and the minerals calcium, phosphorus, and iron.

The West Indian pea is a tall tree with several branches. The compound leaves are approximately 1 ft long, with 12 to 20 pairs of rounded, narrow, oblong leaflets 3 to 4 cm long and approximately 1 cm wide. The leaves are borne at the terminals of branches. The inflorescence is an axillary raceme bearing two or three flowers. The pale pink, red, purple, or white flower buds are 5 to 10 cm long and about 3 cm wide. The pods are 25 to 50 cm long, slender, and cylindrical and have many light brown to red-brown seeds.

The West Indian pea leaf is an excellent source of a number of plant pigments, phytochemical flavonoids (myricetin, quercetin, kaempferol, luteolin, and apigenin), and alpha-tocopherol, which has antioxidant properties associated with lowered risk of cardiovascular diseases and cancers (Ching and Mohamed 2001; Miean and Mohamed 2001). The West Indian pea is also a good source of lutein and zeaxanthin, two pigments important to the cornea, containing 145 and 7.1 mg, respectively, per 100 g DW (Liu et al. 2007). The potential use of West Indian pea tree leaves and leaf extracts for its antibacterial, antibiotic, antimutagenic, anticarcinogenic, and hepatoprotective properties and in the treatment of epileptic fits has also been reported (Kasture et al. 2002; Kusamran 1998; Pari and Uma 2003; Solis 1969).

SOLANACEAE

The Solanaceae is also called the nightshade family, comprising about 85 genera and 2,800 species. Solanaceae plants have been cultivated since ancient times for human food and medicine. The plants are herbs, shrubs, or trees. The calyx and corolla have five lobes, and usually there are five stamens. The fruit may be dry or a fleshy berry and bears numerous seeds in two or four interior divisions.

Eggplant (Solanum melongena; *Solanaceae*)

Eggplant, as it is called by Americans, is also known as Guinea squash in some parts of the southern United States, called *aubergine* by the French and British, *brinjal* in India, and *melanzane* in Italy.

Plants grow to 0.6 to 1.2 m (2-4 ft) and bear characteristic large, lobed, alternate leaves. Flowers are white lavender to pink or blue, with yellow centers. Eggplant fruits vary widely in shape (spherical to cylindrical), size (tiny to enormous), and color (creamy white to dark purple, green, pink, and striped).

Eggplant has been used as a hypocholesterolemic agent in the Asian Subcontinent as well as in the Americas. Although studies with polyphenols extracted from eggplant seem to show them to be beneficial, reducing hypercholesterolemia in animals or having antioxidant activity in experimental models (Noda et al. 2000; Sudheesh et al. 1999), the effect of eggplant or its components in lipid metabolism remains controversial (Noda et al. 2000; Guimarães et al. 2000; Sudheesh et al. 1999).

It has been shown to have antioxidant (Noda et al. 1998), analgesic (Vohora et al. 1984), hypolipidemic (Sudheesh et al. 1997), immunomodulatory (Lee et al. 2001), and anti-inflammatory (Han et al. 2003) effects. The strong hypolipidemic effect combined with an improved ratio of high-density to low-density lipoprotein suggests its use for treating diseases associated with hyperlipidemia such as ischemic heart diseases and arteriosclerosis (Odetola et al. 2004). In 2005 global production was 30,523,867 Mt and the area harvested was 1,765,076 ha (FAOSTAT 2006).

Wolfberry (Lycium chinense; *Solanaceae*)

Wolfberry is grown in China and other parts of Asia and Africa. It is a perennial plant. The leaves are alternate and lanceolate or ovate, the lavender or light purple flowers are bell shaped or tubular, and the ripe berries are bright red. The sweet-tasting berries are believed to enhance the immune system, help eyesight, protect the liver, and improve circulation. In traditional Chinese medicine they are neutral in nature, acting on the liver, lung, and kidney channels and enriching the *yin*. Dried wolfberries, leaves, and shoots are all edible and used as an ingredient in Chinese soups. The fruits and leaves are used as food, tea, and medicine in the Orient. Fruits and leaves are also described as being rich in beta-carotene; vitamins C, B_1, B_2, and E; and the minerals calcium, potassium, and iron (Gross et al. 2006; Young et al. 2005). The leaves contain 17.73 mg alpha-tocopherol per 100 g (Ching and Mohamed 2001). The leaves are capable of reducing the risk of arteriosclerosis, essential arterial hypertension, diabetes, and night blindness (Kim et al. 1997).

LEGUMES

The legume family (Fabaceae or Leguminosae) is second in economic importance to the grass family (Poaceae), and with an estimated 16,000 to 19,000 species, it is ranked third in size after the Orchidaceae and the Compositae (Allen and Allen 1981).

*Pea (***Pisum sativum**; *Fabaceae)*

The green pea or the sweet pea is an important vegetable, known for its crisp, tender, edible seeds, that originated in Asia and spread around the world. In 2005 global green pea production was 9,104,729 Mt and the area harvested was 1,123,991 ha (FAOSTAT 2006).

It is a self-pollinated annual herb grown as a bush or vine. Stems are usually weak and slender, bearing alternate pinnate leaves with one to three pairs of leaflets, which are terminal, branched, tendril, and ovate or elliptic. The inflorescence is a raceme rising from the axil of the leaf. The flowers are white, pink, or purple depending on the variety grown. Pods are swollen or compressed, enclosing 2 to 10 seeds.

The green pea contains 6.7% protein, 0.5% oil, and 13.9% carbohydrate (Şehirali 1988). Immature seedpods are eaten raw or cooked. Young seedpods are sweet and delicious and can be added to salads or lightly cooked or stir-fried. The seeds are also roasted and eaten as a snack and sprouted to use as a salad ingredient. In some parts of the world the tender shoots are cooked and eaten. The seed is contraceptive, fungistatic, and spermicidal (Duke and Ayensu 1985), and the dried powder is used as a poultice for acne. Chopra et al. (1986) described the seed oil as reducing the pregnancy rate in women by 60% and reducing male sperm count by 50% in a two-year period. The pea has great nutritional potential because of its high protein content, and it has been suggested as an alternative protein source to soybean in countries where soybean is not a native crop or for those who have an allergic reaction or intolerance to soybean (Davidsson et al. 2001). The hypocholesterolemic effect of green pea consumption has been demonstrated by Martins et al. (2004), Lasekan et al. (1995), and Dabai et al. (1996).

*French Bean (***Phaseolus vulgaris;** *Fabaceae)*

Green bean (*P. vulgaris* L.), also called string bean, is a commonly consumed vegetable of Asian origin. In 2005 global French bean production was 1,720,125 Mt and the area harvested was 208,366 ha (FAOSTAT 2006).

It is an annual herb grown either as a bush or a vine. Leaves are alternate, green or purple, trifoliolate, ovate, and entire; flowers are racemes borne in the axils. Pods are slender and cylindrical, 10 to 20 cm long and 1 to 2 cm wide, and enclose four to six seeds. French beans are a source of protein, folate, and other important phytochemicals (Wettasinghe et al. 2002).

*Winged Bean (***Psophocarpus tetragonolobus;** *Fabaceae)*

Winged bean is also known as the goa bean or the princess bean. The plant is a vine bearing long leaves and pods that have four angled, leaflike wings running lengthwise to the pods. The seeds are round and green when mature. Young pods are cooked and eaten as a vegetable. The plant is consumed in its entirety because each part has nutritional and medicinal properties (Burkill 1966). The beans have been used as a milk substitute in the treatment of kwashiorkor (Cerny and Adndy 1973).

TUBERS

Tubers are an alternative source of carbohydrates in a number of developing countries and the primary, sometimes the only, source of energy in the diet. A number of tubers have been studied because of their low glycemic index (GI), which makes them a better source of carbohydrates than cereal grains. In a review of vegetable studies for cancer risk reduction by the World Cancer Research Fund (1997) potatoes did not share the beneficial relation with cancer incidence seen with other tubers. Thus, the importance of other minor tubers in the human diet in place of the more commonly used potato as a source of carbohydrates in the diet is acknowledged worldwide.

Yam (**Dioscorea spp.**; *Dioscoreaceae*)

Yam is an important staple in many tropical countries because of the carbohydrate it provides (Akanbi et al. 1996; Omonigho and Ikenebomeh 2000). Several beneficial properties of yams have been recently reported in the literature. When compared with other tropical root crops, yam has been suggested to have nutritional superiority (Wanasundera and Ravindran 1994); its protein content also showed a better amino acid balance for human nutrition (Baquar and Oke 1976; Bradbury 1988; Marcus et al. 1998). The average crude protein content of seven cultivars of *D. alata* tubers was 7.4%; starch, 75.6% to 84.3%; vitamin C, 13.0 to 24.7 mg/100 g FW. Phytic acid content of the yams was low, 58.6 to 198.0 mg/100 g DW. Total oxalate levels in yam tubers ranged from 486 to 781 mg/100 g DW, but the level may not constitute a nutritional concern since 50% to 75% of the oxalates were in water-soluble form (Wanasundera and Ravindran 1994). Other researchers describe the yam as being mainly composed of starch (75%-84% DW), with small amounts of proteins, lipids, and most vitamins, and very rich in minerals (Lasztity et al. 1998; Omonigho 1988). Yam also had a lower GI (62) than potato (71) or cassava (94) (Ramdath et al. 2004).

Some yams are used as medicines in Oriental countries to prevent diarrhea and diabetes (Hsu et al. 1984; Yen 1992). The yam has been used in folk medicine to treat menopausal symptoms. In the diet of postmenopausal women, yams raised the level of sex hormones, lipids, and antioxidants, which might reduce the risk of breast cancer and cardiovascular diseases (Wu et al. 2005). Modern researches have shown that yam extracts can reduce blood sugar (Hikino et al. 1986; Undie and Akubue 1986) and blood lipids (Araghiniknam et al. 1996) and inhibit microbe activity (Hu et al. 1996, 1999; Kelmanson et al. 2000), and they show antioxidant activity (Chan et al. 2004; Farombi et al. 2000). Farombi et al. (2000) demonstrated that brown yam flour contained natural antioxidants and might mitigate damage and disease caused by oxidative components. Chen et al. (2002) reported that a specific type of Taiwanese yam tuber exerted trophic effects in the cecum by mediating luminal fermentation. Fang and Kong (2002) found that yam could enhance serum immunoglobulin G (IgG) concentrations and promote lymphocyte proliferation, thus potentially influencing the body's immune system.

The active components of yam include steroidal sapogenin, glycan, and polyphenol oxidase. Diosgenin, a natural steroidal sapogenin extracted from *Dioscorea* species, reduced serum lipid peroxidation, lowered serum triglyceride and phospholipid levels, and increased high-density lipid levels in older people (Araghiniknam et al. 1996). Discorin, the storage protein of yam tuber, showed scavenging activity, inhibited hypoglycemic effects in animal models (Hikino et al. 1986), and had a hepatoprotective property (Lee et al. 2002). Some bitter types of yam have hypoglycemic effects (McAnuff et al. 2005). Consumption of fresh yam tubers is suggested for prevention of disease in which free radicals are involved, such as cancer and cardiovascular diseases (Bhandari and Kawabata 2004). In 2005 global yam production was 39,904,321 Mt and the area harvested was 4,446,761 ha (FAOSTAT 2006).

*Elephant Yam (***Amorphophallus paeoniif olius, A. campanulatus;** *Araceae)*

Elephant yam, also called sweet yam, is native to tropical Asia. Several of the species belonging to the genus *Amorphophallus* spp. are known for their edible tubers, or corms, eaten as a potato-like vegetable. The name *elephant yam* refers to the size of the corms at harvest and their resemblance to an elephant foot. Elephant yam is called *konjac* in Japan and *mo-yu* in China. It is a common vegetable in several parts of Asia and Africa, cultivated in Malaysia, India, Pakistan, Bangladesh, Indonesia, Sri Lanka, the Philippines, and the Pacific Islands. Elephant yam is an herbaceous plant that grows to a height of 1.2 m (4 ft). The leaves are green, often tinged with pink streaks, large, 30 to 75 cm long, and tripartite, each part subdivided into numerous lobes. The petioles are long (50-80 cm), with red markings. The yellow or brown edible tubers are large, flattened, 20 to 25 cm in diameter and have a central depression. The flowers are spathes, with a strong unpleasant odor, borne on a terminal inflorescence. The black or brown skins of the corms are removed and the tuberous part cooked as a vegetable. The petioles and leaves are also edible; they are cooked and eaten as a vegetable, added to curried dishes, or wrapped around cooked rice or cassava and steamed. The corms are also processed into snack foods such as chips and candied finger food. In Asian traditional medicine the corms are used as carminative, restorative, stomachic, and tonic and also used to treat piles and dysentery. The fresh roots are effective as a stimulant and expectorant and used in India and the Philippines to treat rheumatism. The juice from petioles is fermented and ingested to treat diarrhea. The mashed tubers are believed to relieve constipation. The leaves, corms, and roots are widely used in traditional medical practice in Asia (Chopra et al. 1986; Nepal Department of Medicinal Plants 1993).

The tuber of *A. campanulatus* is useful to treat piles, enlargement of the spleen, abdominal tumors, asthma, abdominal pain, and boils and is often used for acute rheumatism (Yusuf et al. 1994). Yam is used to treat earache, pain, intercostal neuralgia, puerperal fever, and swelling of the throat (Joshi 2000). Elephant yam is a good source of carbohydrate, calcium, phosphorus, and other minerals. The leaves are good sources of carotenoids (Rajyalakshmi et al. 2001). Leaf extracts inhibit aflatoxin production by *Aspergillus flavus* (Prasad et al. 1994), indicating potential for use in the prevention or treatment of tuberculosis. The active principle of elephant yam inhibits in vitro growth of *Mycobacterium tuberculosis* (Gupta and Viswanathan 1956). The tuber is reported to have antiprotease (Prathibha et al. 1995) and analgesic (Shilpi et al. 2005) activities. Glucomannan is a soluble, fermentable, and highly viscous dietary fiber derived from the root of the elephant yam that aids in weight loss, by promoting satiety and fecal energy loss, and improves lipid and lipoprotein parameters and glycemic status (Keithley and Swanson 2005).

*Taro (***Colocasia esulenta;** *Araceae, or Arum)*

Taro is also called dasheen or arvi. It originated in India and parts of Southeast Asia. It is a perennial herb approximately 2 ft tall, with thick shoots from a large corm; slender stolons often branch from the corm, along with offshoot corms. Leaf stalks are long and the blades wide, arrowhead shaped, with the upper surface dark green and velvety. Inflorescence is on a fleshy stalk shorter than leaf petioles; part of the fleshy stalk is enveloped by a long yellow bract. Flowers are tiny and densely crowded on the upper part of the fleshy stalk, with female flowers below and male flowers above. The fruit is a small berry, borne in clusters on the fleshy stalk.

Ramdath et al. (2004) reported a low GI (61), and its consumption as poi, a starchy paste in the Hawaiian diet, is speculated as resulting in lower incidence rates of colorectal cancer (Brown et al. 2005). In 2005 global taro production was 10,586,844 Mt and the area cultivated was 1,841,238 ha (FAOSTAT 2006).

OTHER VEGETABLES

*Breadfruit (***Artocarpus altilis**; *Moraceae)*

Breadfruit is native to the Moluccas, New Guinea, and the Philippines. It is called breadfruit because, when eaten before fully ripe, it feels and tastes like fresh bread. The genus consists of about 50 species, some seeded and some seedless. Breadfruit may be eaten ripe as a fruit or underripe as a vegetable, boiled, or roasted or stuffed with coconut, sugar, or curry mix before roasting. Fully ripe fruits are baked or steamed and seasoned. The steamed fruit is sometimes sliced, rolled in flour, and deep-fried. Breadfruit is also used in soups, puddings, and desserts or candied or pickled. The dried fruit is made into flour and used to make porridge and bread. Soft or overripe breadfruit is used for making chips in Trinidad and Barbados. It is also dried, canned, and sometimes frozen. The seeds from the seeded fruits are boiled, steamed, roasted, and eaten with salt as a snack.

The male inflorescences are candied and eaten, or they are toasted and applied topically to relieve toothaches. The inflorescences also yield a yellow or brown dye. In Trinidad and the Bahamas, a decoction of the breadfruit leaf is taken to lower blood pressure, as a treatment for diabetes, and to relieve asthma. The leaf juice is used to cure earache, and ashes of burned leaves are used to treat skin infections. A powder of roasted leaves is prescribed as a remedy for an enlarged spleen. The latex is also used medicinally to treat broken bones, sprains, bruises, and fungal infections, indicating its probable antimicrobial properties.

The breadfruit tree is a fast-growing tree that can reach a height of 25.5 m (85 ft). The tree has smooth, light-colored bark and a trunk that is as wide as 1.2 m. Many spreading branches form a wide crown at the top. The thick, leathery leaves have a glossy, dark green upper side and a dull green underside with an elevated midrib and veins. The leaves have long petioles, and are broadly ovate and pinnately lobed, with lobes cutting deep to the main midrib. The flowers are tiny, axillary, and monoecious, with the male inflorescence originating first. The male inflorescence is club shaped, up to 5 cm in diameter and 45 cm long. The female inflorescence is rounded or elliptic, is green, and has a prickly head, which develops into the compound fruit (or syncarp). The fruit is composed of more than 1,000 flowers attached to the fruit axis. The tough fruit skin is composed of five- to seven-sided disks with a sharp spine rising from the center. Each of the disks represents the surface of an individual flower. The fruits are globose and are green at first, turning yellowish green, yellow, or yellow brown when ripe. When present in the seeded type, the pale brown seeds with darker stripes are irregularly oval, rounded at one end and pointed at the other. In the center of seedless fruits is a cylindrical or oblong core, bearing flat brown abortive seeds. The fruit is borne singly or in clusters of two or three at the branch tips. The pedicel varies from 2.5 to 12.5 cm (1 to 5 in) in length.

Breadfruit is a culinary vegetable eaten at all stages of maturity, although it is most commonly harvested and consumed when mature and firm, and used as a starchy staple. Its GI (61) is lower than the more widely consumed cassava (Ramdath et al. 2004). The starch content of breadfruit is 15.52 g/100 g FW (Golden and Williams 2001). Some cultivars of breadfruit have a beta-carotene content as high as 868 mg/100 g DW and are a significant source of zeaxanthin and lutein (Englberger et al. 2003a,b). The mature fruits are high in carbohydrates, low in fat and protein, and a good source of minerals and vitamins, especially B vitamins. Breadfruit extractions are effective in controlling parasites (Rasfon 1991). A lectin identified in breadfruit seed is similar to the pectin jacalin from jackfruit that is of great importance and is widely studied by biomedical investigators (Kabir 1998; Remani et al. 1997; Sankaranarayanan et al. 1996; Sujathan et al. 1996; Vijayan et al. 1987). In addition to sharing properties with jacalin, breadfruit seed has the unique property of binding human

IgA (Nedic et al. 2000; Pineau et al. 1990, 1991) and also stimulating in vitro proliferation of human peripheral blood mononuclear cells.

Okra (**Abelmoschus esculentus, Hibiscus esculentus;** *Malvaceae*)

Okra, also known as gumbo and lady's finger, is a tropical vegetable that dates back 2,000 years and may have originated in Africa and Asia. Reports also indicate that okra was known in Egypt in twelfth century AD. The word *okra* was derived from the African name *nkruma*. Okra is cultivated in India, Malaysia, the Philippines, China, Africa, the Caribbean, and other tropical parts of the world. The fruits are usually ready for harvest 70 to 80 days after planting. The fruit pods and leaves are edible. The pods are used in stir-fry, soups, and salads and pickled. The seeds yield edible oil. The pods are dried and stored in African countries and used as a daily vegetable and added to soups for flavor and thickening. Its mucilaginous quality can be disagreeable to some but appealing to others.

The ripe seeds are sometimes used as a substitute for coffee, and the seeds can be dried and powdered for storage. The aromatic compounds from the seeds are used as a perfume in Angola. The flowers, seeds, and roots are all medicinally used in China as a diuretic, as a demulcent, and for difficult labor; in the Malay Peninsula they are used to treat gonorrhea and syphilis. The seeds are used in some Ayurvedic formulas to treat stomach problems. Seeds are valued as an emollient and a demulcent in many African cultures. An infusion of the pods is used to treat dysentery, diarrhea, and acute inflammation and irritation of the stomach, bowels, and kidneys. It is generally believed in India that the consumption of okra improves memory. Okra is also used in folk medicine in Papua New Guinea, Vanuatu, Fiji, New Caledonia, and China for the control of fertility, to ease childbirth, to stimulate lactation, to fight menorrhagia, and to induce abortion (Bourdy and Walter 1992; Proeston 1995).

Okra is a tropical annual growing 0.9 to 1.8 m (3-6 ft) tall. The plant stem is hairy and erect, bearing leaves at alternate nodes. The three- to five-lobed leaves are hairy and cordate. The solitary flowers are self-sterile and axillary, with short pedicels. The five petals are yellow with a red or deep purple base coloring. The fruits are capsules that may be dark green, light green, or red, are rounded, and have about 20 round seeds.

Okra is a good source of vitamins C, A, and B complex, and the minerals iron, phosphorus, and calcium, and it is low in calories (Chavan et al. 1991). Okra has become more common than it was in the past decade and appears in many diet surveys and databases (Block 1994), due to the influence of recent demographic changes in the United States. The antioxidant properties of okra have been demonstrated by Ansari et al. (2005). It has been suggested that okra can help reduce bone loss in conditions of estrogen deficiency and thus provide some protection against osteoporosis (Puel et al. 2005). Several commercial formulations of okra, such as Gastrex, are manufactured from the dried pods and marketed as health food to support immune system, cardiovascular, and gastrointestinal health.

REFERENCES

Adams S. 1992. Purslane eyed as rich food source. Agricultural Research Ag Notes. December.

Agarwal VS, Arya M. 2005. *Alternanthera sessilis* DC. (Amaranthaceae): A new source for the treatment of rickets and marasmus. Ethnobotany 17(1-2): 200-201.

Agudo A, Esteve MG, Pallarés C, Martínez-Ballarín I, Fabregat X, et al. 1997. Vegetable and fruit intake and the risk of lung cancer in women in Barcelona, Spain. European Journal of Cancer 33(8): 1256-1261.

Ahmad N, Hassan MR, Halder H, Bennoor KS. 1999. Effect of *Momordica charantia* (Karolla) extracts on fasting and postprandial serum glucose levels in NIDDM patients. Bangladesh Medical Research Council Bulletin 25: 11-13.

Ahmadiani A, Javan M, Semnanian S, Barat E, Kamalinejad M. 2001. Anti-inflammatory and antipyretic effects of *Trigonella foenum-graecum* leaves extract in the rat. Journal of Ethnopharmacology 75(2-3): 283-286.

Ahmed I, Lakhani MS, Gillett M, John A, Raza H. 2001. Hypotriglyceridemic and hypocholesterolemic effects of anti-diabetic *Momordica charantia* (karela) fruit extract in streptozotocin-induced diabetic rats. Diabetes Research Clinical Practices 51: 155-161.

Akanbi CT, Gureje PO, Adeyemi IA. 1996. Effect of heat-moisture pre-treatment on physical characteristics of dehydrated yam. Journal of Food Engineering 28: 45-54.

Ali L, Khan AK, Mamun MI, Mosihuzzaman M, Nahar N, et al. 1993. Studies on hypoglycemic effects of fruit pulp, seed, and whole plant of *Momordica charantia* on normal and diabetic model rats. Planta Medica 59: 408-412.

Ali M, Thomson M, Afzal M. 2000. Garlic and onions: Their effect on eicosanoid metabolism and its clinical relevance. Prostaglandins, Leukotrienes, and Essential Fatty Acids 62(2): 55-73.

Al-Khaldi SF, Martin SA, Prakash L. 1999. Fermentation of fenugreek fiber, psyllium husk, and wheat bran by *Bacteroides ovatus* V975. Current Microbiology 39(4): 231-232.

Allen ON, Allen EK. 1981. The Leguminosae: A source book of characteristics, uses, and nodulation. University of Wisconsin Press, Madison, WI.

Anila L, Vijayalakshmi NR. 2000. Beneficial effects of flavonoids from *Sesamum indicum, Emblica officinalis* and *Momordica charantia*. Phytotherapy Research 14(8): 592-595.

Ansari NM, Houlihan L, Hussain B, Pieroni A. 2005. Antioxidant activity of five vegetables traditionally consumed by South-Asian migrants in Bradford, Yorkshire, UK. Phytotherapy Research 19(10): 907-911.

Aouadi R, Aouidet A, Elkadhi A, Rayana CB, Jaafoura H, et al. 2000. Effect of fresh garlic (*Allium sativum*) on lipid metabolism in male rats. Nutrition Research 20(2): 273-280.

Araghiniknam M, Chung S, Nelson-White T, Eskelson C, Watson RR. 1996. Antioxidant activity of *dioscorea* and dehydroepiandrosterone (DHEA) in older humans. Life Sciences 59: 47-157.

Aslam M, Anwar F, Nadeem R, Rashid U, Kazi TG, et al. 2005. Mineral composition of Moringa oleifera leaves and pods from different regions of Punjab, Pakistan. Asian Journal of Plant Science 4(4): 417-421.

Augusti KT. 1996. Therapeutic values of onion (*Allium cepa* L.) and garlic (*Allium sativum* L.). Indian Journal of Experimental Biology 34(7): 634-640.

Augusti KT, Sheela CG. 1996. Antiperoxide effect of S-allyl cysteine sulfoxide, an insulin secretagogue in diabetic rats. Experientia 52: 115-119.

Awad AB, Begdache LA, Carol SF. 2000. Effect of sterols and fatty acids on growth and triglyceride accumulation in 3T3-L1 cells. Journal of Nutritional Biochemistry 11(3): 153-158.

Bajpai M, Pande A, Tewari SK, Prakash D. 2005. Phenolic contents and antioxidant activity of some food and medicinal plants. International Journal of Food Sciences and Nutrition 56(4): 287-291.

Banerjee AK, Jain M, Dubey V. 1992. Constituents of *Basella rubra* seed oil. Fitoterapia 63(4): 377.

Baquar SR, Oke OL. 1976. Protein in Nigerian yams (*Dioscorea* spp.). Nutrient Reports International 14: 237-248.

Bari ML, Kusunoki H, Furukawa H, Ikeda H, Isshiki K, et al. 1999. Inhibition of growth of *Escherichia coli* O157: H7 in fresh radish (*Raphanus sativus* L.) sprout production by calcinated calcium. Journal of Food Protection 62(2): 128-132.

Basch E, Gabardi S, Ulbricht C. 2003. Bitter melon (*Momordica charantia*): A review of efficacy and safety. American Journal of Health and Systemic Pharmacology 65: 356-359.

Battelli MG, Polito L, Bolognesi A, Lafleur L, Fradet Y, et al. 1996. Toxicity of ribosome-inactivating proteins-containing immunotoxins to a human bladder carcinoma cell line. International Journal of Cancer 68: 485-490.

Baumann J, Bruchhausen FV, Wurm G. 1980. Flavonoid and related compounds as inhibitors of arachidonic acid peroxidation. Prostaglandin 20: 627-639.

Beatty ER, O'Reilly JD, England TG, McAnlis GT, Young IS, et al. 2000. Effect of dietary quercetin on oxidative DNA damage in healthy human subjects. British Journal of Nutrition 84(6): 919-925.

Bhandari MR, Kawabata J. 2004. Organic acid, phenolic content and antioxidant activity of wild yam (*Dioscorea* spp.) tubers. Food Chemistry 88(2): 163-168.

Bhaskarachary K, Sankar Rao DS, Deosthale YG, Reddy V. 1995. Carotene content of some common and less familiar foods of plant origin. Food Chemistry 54(2): 189-193.

Bhattacharjee LI, Mudambi SR, Bhushan B, Patankar MV. 1994. Provitamin A content of selected Indian foods. Journal of Food Science and Technology 31(3): 249-251.

Biswas AR, Ramaswamy S, Bapna JS. 1991. Analgesic effect of *Momordica charantia* seed extract in mice and rats. Journal of Ethnopharmacology 31: 115-118.

Bjorneboe A, Bjorneboe G, Drevon C. 1990. Absorption, transport and distribution of vitamin E. Journal of Nutrition 120: 233-242.

Block G. 1994. Nutrient sources of provitamin A carotenoids in the American diet. American Journal of Epidemiology 139(3): 290-293.

Bordia T, Mohammed N, Thomson M, Ali M. 1996. An evaluation of garlic and onion as antithrombotic agents. Prostaglandins, Leukotrienes, and Essential Fatty Acids 54(3): 183-186.

Bordia A, Verma SK, Srivastava KC. 1997. Effect of ginger (*Zingiber officinale Rosc*) and fenugreek (*Trigonella foenumgraecum* L) on blood lipids, blood sugar, and platelet aggregation in patients with coronary artery disease. Prostaglandins, Leukotrienes, and Essential Fatty Acids 56: 379-384.

Bordia A, Verma SK, Srivastava KC. 1998. Effect of garlic (*Allium sativum*) on blood lipids, blood sugar, fibrinogen and fibrinolytic activity in patients with coronary artery disease. Prostaglandins, Leukotrienes, and Essential Fatty Acids 58(4): 257-263.

Borek C. 2001. Antioxidant health effects of aged garlic extract. Journal of Nutrition 131(3 suppl.): 1010S-1015S.

Bosworth SC, Hoveland CS, Buchanan GA, Anthony WB. 1980. Forage quality of selected warm-season weed species. Agronomy Journal 72: 1050-1054.

Bourdy G, Walter A. 1992. Maternity and medicinal plants in Vanuatu. I. The cycle of reproduction. Journal of Ethnopharmacology 37: 179-196.

Bradbury JH. 1988. The chemical composition of tropical root crops. ASEAN Food Journal 4: 3-13.

Brahmachari HD, Augusti KT. 1961. Hypoglycaemic agent from onions. Journal of Pharmacy and Pharmacology 13: 128.

Brahmachari HD, Augusti KT. 1962. Orally effective hypoglycaemic agents from plants. Journal of Pharmacy and Pharmacology 14: 254-255.

Bravo L. 1998. Polyphenols: Chemistry, dietary sources, metabolism, and nutritional significance. Nutritional Reviews 56: 317-333.

Brown AC, Reitzenstein, JE, Liu J, Jadus MR. 2005. The anti-cancer effects of poi (*Colocasia esculenta*) on colonic adenocarcinoma cells *in vitro*. Phytotherapy Research 19(9): 767-771.

Burkill IH. 1966. A dictionary of the economic products of the Malay Peninsula, Vol. 1. The Ministry of Agriculture and Cooperatives, Kuala Lumpur, Malaysia.

Caceres A, Cabrera O, Morales O, Mollinedo P, Mendia P. 1991. Pharmacological properties of *Moringa oleifera*. 1: Preliminary screening for antimicrobial activity. Journal of Ethnopharmacology 33(3): 213-216.

Caceres A, Saravia A, Rizzo S, Zabala L, De Leon E, et al. 1992. Pharmacologic properties of *Moringa oleifera*. 2: Screening for antispasmodic, antiinflammatory and diuretic activity. Journal of Ethnopharmacology 36(3): 233-237.

Cassidy A, Hanley B, Lamuela-Raventos RM. 2000. Isoflavones, lignans and stilbenes: Origins, metabolism and potential importance to human health. Journal of the Science of Food and Agriculture 80: 1044-1062.

Castro M. 1990. The complete homeopathy handbook. Macmillan, London.

Cerny K, Adndy HA. 1973. The winged bean (*Psophocarpus palustri* Desv.) in the treatment of kwashiorkor. British Journal of Nutrition 29: 105-107.

Chan YC, Hsu CK, Wang MF, Su TY. 2004. A diet containing yam reduces the cognitive deterioration and brain lipid peroxidation in mice with senescence accelerated. International Journal of Food Science and Technology 39(1): 99-107.

Chan C, Leung I, Lam KW, Tso MOM. 1998. The occurrence of retinol and carotenoids in human subretinal fluid. Current Eye Research 17: 890-895.

Chan WY, Ng TB, Yeung HW, So KC. 1985. The inhibitory effects of beta-momorcharin on endometrial cells in the mouse. Contraception 31: 83-90.

Chan WY, Ng TB, Yeung HW. 1984. The termination of early pregnancy in the mouse by beta-momorcharin. Contraception 29: 91-100.

Chan WY, Ng TB, Yeung HW, Tam PP, Choi HL. 1986. Effects of momorcharins on the mouse embryo at the early organogenesis stage. Contraception 34: 537-544.

Chanwitheesuk A, Terrawutgulrag A, Rakariyatham N. 2005. Screening of antioxidant activity and antioxidant compounds of some edible plants of Thailand. Food Chemistry 92(3): 491-497.

Chavan UD, Adsule RN, Kachare DP. 1991. Chemical composition and nutritional quality of some promising cultivars of okra. Journal of Maharashtra Agricultural Universities 16(2): 287-288.

Chen HL, Chang CT, Wang CH. 2002. The trophic effects of Taiwanese Yam (*Dioscorea* spp.) tubers on the epithelial cells in the lower gut of mice. Health Food Society of Taiwan, Taipei, Taiwan. p. 68.

Chen JH, Chen HI, Wang JS, Tsai SJ, Jen CJ. 2000. Effects of Welsh onion extracts on human platelet function *in vitro*. Life Sciences 66(17): 1571-1579.

Chevallier A. 1996. The encyclopedia of medicinal plants. Dorling Kindersley, London.

Ching LS, Mohamed S. 2001. Alpha-tocopherol content in 62 edible tropical plants. Journal of Agricultural and Food Chemistry 49(6): 3101-3105.

Choi HY, Park YK, Kim DI, Kim YH, Yoon GB, et al. 2001. Donguibogam. Yeogang Press, Seoul. p. 374-2022.

Choi J, Lee KT, Jung H, Park HS, Park HJ. 2002. Anti-rheumatoid arthritis effect of the *Kochia scoparia* fruits and activity comparison of momordin Ic, its prosapogenin and sapogenin. Archives of Pharmacological Research 25: 336-342.

Chopra RN, Nayar SL, Chopra IC. 1986. Glossary of Indian medicinal plants. Council of Scientific and Industrial Research, New Delhi.

Choudhary D, Chandra D, Choudhary S, Kale RK. 2001. Modulation of glyoxalase, glutathione S-transferase and antioxidant enzymes in the liver, spleen and erythrocytes of mice by dietary administration of fenugreek seeds. Food and Chemical Toxicology 39(10): 989-997.

Chu YH, Chang CL, Hsu HF. 2000. Flavonoid content of several vegetables and their antioxidant activity. Journal of the Science of Food and Agriculture 80(5): 561-566.

Claflin AJ, Vesely DL, Hudson JL, Bagwell CB, Lehotay DC, et al. 1978. Inhibition of growth and guanylate cyclase activity of an undifferentiated prostate adenocarcinoma by an extract of the balsam pear (*Momordica charantia* abbreviata). Proceedings of the National Academy of Sciences of the United States of America 75: 989-993.

Coquillat M. 1951. Sur les plantes plus communes a la surface du globe. Bulletin Mensuel de la Société Linnéenne de Lyon. p. 165-170.

Corlett JL, Clegg MS, Keen CL, Grivetti LE. 2002. Mineral content of culinary and medicinal plants cultivated by Hmong refugees living in Sacramento, California. International Journal of Food Science and Nutrition 53(2): 117-128.

Cunnick JE, Sakamoto K, Chapes SK, Fortner GW, Takemoto DJ. 1990. Induction of tumor cytotoxic immune cells using a protein from the bitter melon (*Momordica charantia*). Cellular Immunology 126: 278-289.

Dabai FD, Walker AF, Sambrook IE, Welch VA, Owen RW, et al. 1996. Comparative effects on blood lipids and faecal steroids of five legume species incorporated into a semi-purified, hypercholesterolaemic rat diet. British Journal of Nutrition 75(4):557-571.

Dagnelie G, Zorge IS, McDonald TM. 2000. Lutein improves visual function in some patients with retinal degeneration, a pilot study via the Internet. Optometry 71: 147-164.

Dahot MU. 1988. Vitamin contents of the flowers and seeds of *Moringa oleifera*. Pakistan Journal of Biochemistry 21(1-2): 21-24.

Das BR, Kurup PA, Rao PL, Narasimha R. 1957. Antibiotic principle from *Moringa pterygosperma*. VII. Antibacterial activity and chemical structure of compounds related to pterygospermin. Indian Journal of Medical Research 45: 191-196.

Davidsson L, Dimitriou T, Walczyk T, Hurrel RF. 2001. Iron absorption from experimental infant formulas based on pea (*Pisum sativum*)-protein isolate: The effect of phytic acid and ascorbic acid. British Journal of Nutrition 85: 59-63.

Daxenbichler ME, Spencer FG, Carlson DG, Rose GB, Brinker AM, et al. 1991. Glucosinolate composition of seeds from 297 species of wild plants. Phytochemistry 30: 2623-2638.

De Vogel J, Jonker-Termont DS, van Lieshout EM, Katan MB, van der Meer R. 2005. Green vegetables, red meat and colon cancer: Chlorophyll prevents the cytotoxic and hyperproliferative effects of haem in rat colon. Carcinogenesis 26(2): 387-393.

De Whalley CV, Rankin SM, Hoult JRS, Jessup W, Leake DS. 1990. Flavonoids inhibit the oxidative modification of low-density lipoprotein by macrophages. Biochemical Pharmacology 39: 1743-1750.

Dhar P, Ghosh S, Bhattacharyya DK. 1999. Dietary effects of conjugated octadecatrienoic fatty acid (9 *cis*, 11 *trans*, 13 *trans*) levels on blood lipids and nonenzymatic *in vitro* lipid peroxidation in rats. Lipids 34: 109-114.

Dirsch VM, Gerbes AL, Vollmar AM. 1998. Ajoene, a compound of garlic, induces apoptosis in human promyeloleukemic cells, accompanied by generation of reactive oxygen species and activation of nuclear factor kappa. Molecular Pharmacology 53(3): 402-407.

Dorant E, van den Brandt PA, Goldbohm RA, Hermus RJ, Sturmans F. 1993. Garlic and its significance for the prevention of cancer in humans: A critical view. British Journal of Cancer 67(3): 424-429.

Duke JA. 1983a. *Moringa oleifera* Lam. In: Handbook of energy crops, http://www.hort.purdue.edu/newcrop/duke_energy/Moringa_oleifera.html. Accessed July 10, 2006.

Duke JA. 1983b. *Sesbania grandiflora* (L.) Pers. In: Handbook of energy crops, http://www.hort.purdue.edu/newcrop/duke_energy/Sesbania_grandiflora.html. Accessed July 10, 2006.

Duke JA, Ayensu ES. 1985. Medicinal plants of China. Reference Publications, Inc., Algonac, MI.

Duke JA, Wain KK. 1981. Medicinal plants of the world. Computer index with more than 85,000 entries. 3 vols. Food and Agricultural Organization, United Nations, Rome.

Durust N, Sumengen D, Durust Y. 1997. Ascorbic acid and element contents of foods of Trabzon (Turkey). Journal of Agricultural and Food Chemistry 45(6): 2085-2087.

Dwivedi SN, Dwivedi S, Patel PC. 2005. Herbal remedies used by the rural people in the treatment of joint diseases. Ethnobotany 17(1-2): 193-196.

Eilert U, Wolters B, Nahrstedt A. 1981. The antibiotic principle of seeds of *Moringa oleifera* and *Moringa stenopetala*. Journal of Medicinal Plant Research 42: 55-61.

Elless MP, Blaylock MJ, Huang JW, Gussman CD. 2000. Plants as a natural source of concentrated mineral nutritional supplements. Food Chemistry 71(2): 181-188.

Englberger L, Aalbersberg W, Ravi P, Bonnin E, Marks GC, et al. 2003a. Further analyses on Micronesian banana, taro, breadfruit and other foods for provitamin A carotenoids and minerals. Journal of Food Composition and Analysis 16(2): 219-236.

Englberger L, Schierle J, Marks GC, Fitzgerald MH. 2003b. Micronesian banana, taro, and other foods: newly recognized sources of provitamin A and other carotenoids. Journal of Food Composition and Analysis 16(1), 3-19.

Fahey JW, Haristoy X, Dolan PM, Kensler TW, Scholtus I, et al. 2002. Sulforaphane inhibits extracellular, intracellular, and antibiotic-resistant strains of *Helicobacter pylori* and prevents benzo[a]pyrene-induced stomach tumors. Proceedings of the

National Academy of Sciences of the United States of America 99(11): 7610-7615.

Faizi S, Siddiqui BS, Saleem R, Aftab K, Shaheen F, et al. 1998. Hypotensive constituents from the pods of *Moringa oleifera*. Planta Medica 64(3): 225-228.

Fang MC, Kong RL. 2002. Immuno-bioactive of yam crude polysaccharide and yam yogurt in cell culture and mice. Health Food Society of Taiwan, Taipei, Taiwan. p. 66.

FAOSTAT (Statistics Division of the Food and Agriculture Organization of the United Nations). 2006. http://faostat.fao.org/. Accessed February 2006.

Farombi EO, Britton G, Emerole GO. 2000. Evaluation of the antioxidant activity and partial characterization of extracts from browned yam flour diet. Food Research International 33: 493-499.

Feng PC, Haynes LJ, Magnus KE. 1961. High concentration of (-)-noradrenaline in *Portulaca oleracea* L. Nature 191: 1108.

Fleischauer AT, Poole C, Arab L. 2001. Garlic consumption and cancer prevention: Meta-analyses of colorectal and stomach cancers. American Journal of Clinical Nutrition 72(4): 1047-1052.

Fletcher R. 1998. *Sauropus androgynus* (sweet leaf bush). Australian New Crops Newsletter 9. Available at: http://www.newcrops.uq.edu.au/newslett/ncnl9191.htm. Accessed January 31, 2004.

Foa-Tomasi L, Campadelli-Fiume G, Barbieri L, Stirpe F. 1982. Effect of ribosome-inactivating proteins on virus-infected cells. Inhibition of virus multiplication and of protein synthesis. Archives of Virology 71: 323-332.

Frei B, Higdon JV. 2003. Antioxidant activity of tea polyphenols in vivo: Evidence from animal studies. Journal of Nutrition 133(10): 3275S-3284S.

Freiberger CE, Vanderjagt DJ, Pastuszyn A, Glew RS, Mounkaila G, et al. 1998. Nutrient content of the edible leaves of seven wild plants from Niger. Plant Foods for Human Nutrition 53: 57-69.

Fukushima S, Takada N, Hori T, Wanibuchi H. 1997. Cancer prevention by organosulfur compounds from garlic and onion. Journal of Cellular Biochemistry (Suppl. 27): 100-105.

Ganguly C, De S, Das S. 2000. Prevention of carcinogen-induced mouse skin papilloma by whole fruit aqueous extract of *Momordica charantia*. European Journal of Cancer Prevention 9: 283-288.

Gey KF, Puska P. 1989. Plasma vitamins E and A inversely correlated to mortality from ischemic heart disease in cross-cultural epidemiology. Annals of the New York Academy of Sciences 570: 268-282.

Ghahfarokhi MS, Shokoohamiri MR, Amirrajab N, Moghadasi B, Ghajari A, et al. 2006. *In vitro* antifungal activities of *Allium cepa*, *Allium sativum* and ketoconazole against some pathogenic yeasts and dermatophytes. Fitoterapia 77(4): 321-323.

Ghasi S, Nwobodo E, Ofili JO. 2000. Hypocholesterolemic effects of crude extract of leaf of *Moringa oleifera* Lam. in high-fat diet fed wistar rats. Journal of Ethnopharmacology 69: 21-25.

Giri, J, Bhuvaneswari V, Rajeswari D. 1984. Changes in the nutritive value of chekkurmenis at different stages of growth. Indian Journal of Nutrition and Dietetics 21(11): 419-423.

Giron LM, Freire V, Alonzo A, Caceres A. 1991. Ethnobotanical survey of the medicinal flora used by the Caribs of Guatemala. Journal of Ethnopharmacology 34: 73-187.

Golden KD, Williams OJ. 2001. Amino acid, fatty acid, and carbohydrate content of *Artocarpus altilis* (breadfruit). Journal of Chromatography Science 39(6): 243-250.

Goldman IL, Kopelberg M, Debaene JE, Schwartz BS. 1996. Antiplatelet activity in onion (*Allium cepa*) is sulfur dependent. Thrombosis and Haemostasis 76(3): 450-452.

Granado F, Olmedilla B, Blanco I. 2003. Nutritional and clinical relevance of lutein in human health. British Journal of Nutrition 90: 487-502.

Grieve M. 1982. A modern herbal. Vol. 1. Dover Publications, New York. p. 342-345.

Gross PM, Zhang X and Zhang R (2006). Wolfberry: Nature's bounty of nutrition and health. BookSurge Publishing, Charleston, SC.

Grover JK, Rathi SS, Vats V. 2000. Preliminary study of fresh juice of *Benincasa hispida* on morphine addiction in mice. Fitoterapia 71: 707-709.

Grover JK, Rathi SS, Vats V. 2002. Amelioration of experimental diabetic neuropathy and gastropathy in rats following oral administration of plant (*Eugenia jambolana, Mucuna pruriens* and *Tinospora cordifolia*) extracts. Indian Journal of Experimental Biology 40: 273-276.

Grover JK, Vats V, Rathi SS, Dawar R. 2001. Traditional Indian anti-diabetic plants attenuate progression of renal damage in streptozotocin induced diabetic mice. Journal of Ethnopharmacology 76: 233-238.

Guevara AP, Vargas C, Sakurai H, Fujiwara Y, Hashimoto K, et al. 1999. An anti-tumor promoter from *Moringa oleifera* Lam. Mutation Research 440: 181-188.

Guimarães RP, Galvão PMA, Batista MC, Azevedo SG, Oliveira DR, et al. 2000. Eggplant (*Solanum melongena*) infusion has a modest and transitory effect on hypercholesterolemic subjects. Brazilian Journal of Medical and Biological Research 33: 1027-1036.

Gupta KC, Viswanathan R. 1956. Antituberculous substances from plants. Antibiotics and Chemotherapy 6: 194-195.

Gurbuz I, Akyuz C, Yesilada E, Sener B. 2000. Anti-ulcerogenic effect of *Momordica charantia* L. fruits on various ulcer models in rats. Journal of Ethnopharmacology 7: 77-82.

Hafeez BB, Haque R, Parvez S, Pandey S, Sayeed I, et al. 2003. Immunomodulatory effects of fenugreek (*Trigonella foenum graecum* L.) extract in mice. International Immunopharmacology 3(2): 257-265.

Han SW, Tae J, Kim JA, Kim DK, Seo GS, et al. 2003. The aqueous extract of *Solanum melongena* inhibits PAR2 agonist-induced inflammation. Clinica Chimica Acta 328(1-2): 39-44.

Harit M, Rathee PS. 1995. Antibacterial activity of the unsaponifiable fraction of the fixed oils of *Trichosanthes* seeds. Asian Journal of Chemistry 7(4): 909-911.

Harper AE. 1999. Defining the essentiality of nutrients. In: Shils ME, Olson JA, Shike M, Ross AC, editors. Modern nutrition in health and disease. 9th ed. Williams and Wilkins, Baltimore, MD. p. 3-10.

Hartwell JL. 1967-1971. Plants used against cancer. A survey. Lloydia 30-34.

Hasanah M, Yuliani S, Risfaheri A. 1999. Prospect of katuk plant (*Sauropus androgynus*) to be developed as a medicinal product. Indonesian Agricultural Research and Development Journal 21(4): 59-64.

Haskell MJ, Jamil KM, Hassan F, Peerson JM, Hossain MI, et al. 2004. Daily consumption of Indian spinach (*Basella alba*) or sweet potatoes has a positive effect on total-body vitamin A stores in Bangladeshi men. American Journal of Clinical Nutrition 80(3): 705-714.

Hayashi K, Takehisa T, Hamato N, Takano R, Hara S, et al. 1994. Inhibition of serine proteases of the blood coagulation system by squash family protease inhibitors. Journal of Biochemistry 116: 1013-1018.

Hebbar SS, Harsha VH, Shripathi V, Hegde GR. 2004. Ethnomedicine of Dharwad district in Karnataka, India: Plants used in oral health care. Journal of Ethnopharmacology 94(2-3): 261-266.

Hecht SS. 1999. DNA adduct formation from tobacco-specific *N*-nitrosamines. Mutation Research/Fundamental and Molecular Mechanisms of Mutagenesis 424(1-2): 127-142.

Helen A, Rajasree CR, Krishnakumar K, Augusti KT, Vijayammal PL. 1999. Antioxidant role of oils isolated from garlic (*Allium sativum* Linn) and onion (*Allium cepa* Linn) on nicotine-induced lipid peroxidation. Veterinary and Human Toxicology 41(5): 316-319.

Hemalatha G, Sundharaiya K, Ponnuswamy V. 1999. Comparative analysis of nutritive value in some leafy vegetables. South Indian Horticulture 47(1-6): 295.

Hertog MGL. 1995. Flavonoid intake and long-term risk of coronary heart disease and cancer in the 7 countries study. Archives of Internal Medicine 155(11): 1184-1195.

Hikino H, Konno C, Takahashi M, Murakami M, Kato Y, et al. 1986. Isolation and hypoglycemic activity of dioscorans A, B, C, D, E, and F; glycans of *Dioscorea japonica* rhizophors. Planta Medica 3: 168-171.

Hollman PCH, Van Trijp JMP, Buysman MNCP, van der Gaag MS, Mengelers MJ, et al. 1997. Relative bioavailability of the antioxidant flavonoid quercetin from various foods in man. FEBS Letters 418: 152-156.

Holm LG, Plucknett DL, Pancho JV, Herberger JP. 1977. The world's worst weeds: Distribution and biology. Vol. 7. University of Hawaii Press, Honolulu. p. 78-83.

Hossain MZ, Shibib BA, Rahman R. 1992. Hypoglycemic effects of *Coccinia indica*: Inhibition of key gluconeogenic enzyme, glucose-6-phosphatase. Indian Journal of Experimental Biology 30(5): 418-420.

Hsu HY, Chen YP, Hsu SJ, Chen JJ, Chang SC. 1984. Essentials of materia medica. New Medicine and Pharmacy Press, Taipei, Taiwan.

Hu K, Dong A, Yao X, Kobayashi H, Iwasaki S, et al. 1996. Antineopastic agents: I. Three spirostanol glycosides from rhizomes of *Dioscorea collettii* var. hypoglauca. Planta Medica 62: 573-575.

Hu K, Dong A, Yao X, Kobayashi H, Iwasaki S, et al. 1999. A new pregnane from *Dioscorea collettii* var. hypoglauca. Journal of Natural Products 62: 299-301.

Hu L, Tang Z, Cao Y, Ye Q, Huang X. 1989. Preliminary report on natural *Basella* red pigment. Tianran Chanwu Yanjiu Yu Kaifa 1(1): 67-70.

Husain J, Tickle IJ, Wood SP. 1994. Crystal structure of momordin, a type I ribosome inactivating protein from the seeds of *Momordica charantia*. FEBS Letters 342: 154-158.

Inoi R, Anabuki K, Iwasaki H, Akiya M, Inoue N. 2001. Analysis of *Trichosanthes anguina* fruit. Suginami-ku Eisei Shikensho Nenpo 19: 56-57.

Ip C, Lisk DJ, Thompson HJ. 1996. Selenium-enriched garlic inhibits the early stage but not the late stage of mammary carcinogenesis. Carcinogenesis 17(9): 1979-1982.

Jahn SAA. 1981. Traditional water purification in tropical developing countries: Existing methods and potential applications. Manual No. 117. Pub. GTZ, Eschborn, Germany.

Jahn SAA, Musnad HA, Burgstaller H. 1986. The tree that purifies water: Cultivating multipurpose Moringaceae in the Sudan. Unasylva 38: 23-28.

Janot MM, Laurin J. 1930. Hypoglycaemic action of the bulbs of *Allium cepa* Linn. Comptes Rendus de l'Académie des Sciences de Paris 191: 1098.

Jayasooriya AP, Sakono M, Yukizaki C, Kawano M, Yamamoto K, et al. 2000. Effects of *Momordica charantia* powder on serum glucose levels and various lipid parameters in rats fed with cholesterol-free and cholesterol-enriched diets. Journal of Ethnopharmacology 72: 331-336.

Jayaweera DMA. 1981. Medicinal plants used in Ceylon: Part III, Royal Botanic Garden, Peradeniya. National Science Council of Sri Lanka, Colombo, Sri Lanka. p. 255.

John S, Chellappa AR. 2005. Hypoglycemic effect of *Moringa oleifera* (drumstick) leaf powder on human diabetic subjects and albino rats. Indian Journal of Nutrition and Dietetics 42(1): 22-29.

Joshi SG. 2000. Medicinal plants. IBH Publishing, Calcutta, India. p. 33.

Kabir S. 1998. Jacalin: A jackfruit (*Artocarpus heterophyllus*) seed-derived lectin of versatile applications in immunobiological research. Journal of Immunological Methods 212(2): 193-211.

Kamble SM, Kamlakar PL, Vaidya S, Bambole VD. 1998. Influence of *Coccinia indica* on certain enzymes in glycolytic and lipolytic pathway in human diabetes. Indian Journal of Medical Sciences 52(4): 143-146.

Kapoor LD. 1990. Handbook of Ayurvedic medicinal plants. CRC Press, Boca Raton, FL.

Kasture VS, Deshmukh VK, Chopde CT. 2002. Anxiolytic and anticonvulsive activity of Sesbania grandiflora leaves in experimental animals. Phytotherapy Research 6(5): 455-460.

Kawabata J, Kasai T. 2000. Cruciferous plants or their extracts as alpha-glucosidase inhibitors for treatment of obesity and diabetes. Japanese Kokai Tokkyo Koho. p. 3.

Kays SJ, Dias JCS. 1995. Common names of commercially cultivated vegetables of the world in 15 languages. Economic Botany 40(2): 115-152.

Keithley J, Swanson B. 2005. Glucomannan and obesity: A critical review. Alternative Therapies in Health and Medicine 11(6): 30-34.

Kelmanson JE, Jager AK, Van Staden J. 2000. Zulu medicinal plants with antibacterial activity. Journal of Ethnopharmacology 69: 241-246.

Kendler BS. 1987. Garlic (*Allium sativum*) and onion (*Allium cepa*): A review of their relationship to cardiovascular disease. Preventive Medicine 16(5): 670-685.

Khan BA, Abraham A, Leelamma S. 1995. Hypoglycemic action of *Murraya koenigii* (curry leaf) and *Brassica juncea* (mustard): Mechanism of action. Indian Journal of Biochemistry and Biophysics 32(2): 106-108.

Khan BA, Abraham A, Leelamma S. 1996a. Biochemical response in rats to the addition of curry leaf (*Murraya koenigii*) and mustard seeds (*Brassica juncea*) to the diet. Plant Foods for Human Nutrition 49(4): 295-299.

Khan BA, Abraham A, Leelamma S. 1996b. Role of *Murraya koenigii* (curry leaf) and *Brassica juncea* (mustard) in lipid peroxidation. Indian Journal of Physiology and Pharmacology 40(2): 155-158.

Khan BA, Abraham A, Leelamma S. 1996c. *Murraya koenigii* and *Brassica juncea*: Alterations on lipid profile in 1-2 dimethyl hydrazine induced colon carcinogenesis. Investigational New Drugs 14(4): 365-369.

Khan BA, Abraham A, Leelamma S. 1997. Anti-oxidant effects of curry leaf, *Murraya koenigii* and mustard seeds, *Brassica juncea* in rats fed with high fat diet. Indian Journal of Experimental Biology 35(2): 148-150.

Khan MR. 1998. *Momordica charantia* and *Allium sativum*: Broad-spectrum antibacterial activity. Korean Journal of Pharmacognosy 29: 155-158.

Khanum F, Swamy MS, Krishna KRS, Santhanam K, Viswanathan KR. 2000. Dietary fiber content of commonly fresh and cooked vegetables consumed in India. Plant Foods for Human Nutrition 55(3): 207-218.

Kim JH. 1997. Anti-bacterial action of onion (*Allium cepa* L.) extracts against oral pathogenic bacteria. Journal of Nihon University School of Dentistry 39(3): 136-141.

Kim SY, Lee KH, Chang KS, Bock JY, Jung MY. 1997. Taste and flavor compounds in box thorn (*Lycium chinense* Miller) leaves. Food Chemistry 58(4): 297-303.

Kim YJ, Shin MG. 1999. Mucolytic effects of various parts of Fructus *Benincasa* extracts in the rat trachea. Journal of Korean Medical Society 20: 165-176.

Kirtikar KR, Basu BD. 1935. Indian medicinal plants. 2nd ed. Vols. 1 and 3. Bishen Singh Mahendra Pal Singh, Dehra Dun.

Knekt P, Isotupa S, Rissanen H, Heliovaara M, Jarvinen R, et al. 2000. Quercetin intake and the incidence of cerebrovascular disease. European Journal of Clinical Nutrition 54: 415-417.

Kohler I, Jenett-Siems K, Siems K, Hernandez MA, Ibarra RA, et al. 2002. In vitro antiplasmodial investigation of medicinal plants from El Salvador. Zeitschrift fur Naturforschung [section C] 57: 277-281.

Kostova AT, Jenkins SN, Fahey JW, Ye L, Wehage SL, et al. 2005. Protection against UV-light-induced skin carcinogenesis in SKH-1 high-risk mice by sulforaphane-containing broccoli sprout extracts. Cancer Letters 28; 240(2):243-252.

Krinsky NL, Landrum JT, Bone RA. 2003. Biological mechanism of the protective role of lutein and zeaxanthin in the eye. Annual Review of Nutrition 23: 171-201.

Kumar A. 2004. Influence of radish consumption on urinary calcium oxalate excretion. Nepal Medical College Journal 6(1): 41-44.
Kumar GP, Sudheesh S, Ushakumari B, Valsa AK, Vijayakumar S, et al. 1997. A comparative study on the hypolipidemic activity of eleven different pectins. Journal of Food Science and Technology 34(2): 103-107.
Kumar GP, Sudheesh S, Vijayalakshmi NR. 1993. Hypoglycaemic effect of *Coccinia indica*: Mechanism of action. Planta Medica 59(4): 330-332.
Kumaresan PT, Kumar PRAV, Rajasekaran A, Arivukarasu R, Balasubramanian S. 2001. Anti-bacterial activity of aerial parts of *Alternanthera sessilis*. Indian Journal of Natural Products 17(1): 23-24.
Kumari K, Mathew BC, Augusti KT. 1995. Antidiabetic and hypolipidemic effects of S-methyl cysteine sulfoxide isolated from *Allium cepa* Linn. Indian Journal of Biochemistry and Biophysics 32(1): 49-54.
Kurilich AC, Tsau GJ, Brown A, Howard L, Klein BP, et al. 1999. Carotene, tocopherol, and ascorbate content in subspecies of *Brassica oleracea*. Journal of Agricultural and Food Chemistry 47(4): 1576-1581.
Kusamran WR, Tepsuwan A, Kupradinun P. 1998. Antimutagenic and anti-carcinogenic potentials of some Thai vegetables. Mutation Research 402(1-2): 247-258.
Lakshminarayana R, Raju M, Krishnakantha TP, Baskaran V. 2005. Determination of major carotenoids in a few Indian leafy vegetables by high-performance liquid chromatography. Journal of Agricultural and Food Chemistry 53(8): 2838-2842.
Lal J, Chandra S, Raviprakash V, Sabir M. 1976. *In vitro* anthelmintic action of some indigenous medicinal plants on *Ascardia galli* worms. Indian Journal of Physiology and Pharmacology 20: 64-68.
Lamm DL, Riggs DR. 2000. The potential application of *Allium sativum* (garlic) for the treatment of bladder cancer. Urologic Clinics of North America 27(1): 157-162.
Lamm DL, Riggs DR. 2001. Enhanced immunocompetence by garlic: Role in bladder cancer and other malignancies. Journal of Nutrition 131(Suppl. 3): 1067S-1070S.
Lampe JW, King IB, Li S, Grate MT, Barale KV, et al. 2000. *Brassica* vegetables increase and apiaceous vegetables decrease cytochrome P450 1A2 activity in humans: Changes in caffeine metabolite ratios in response to controlled vegetable diets. Carcinogenesis 21(6): 1157-1162.
Landrum JT Bone RA, 2001. Lutein; zeaxanthin; and the macular pigment. Archives of Biochemistry and Biophysics 385: 28-40.
Lans C, Brown G. 1998. Observations on ethnoveterinary medicines in Trinidad and Tobago. Preventive Veterinary Medicine 35: 125-142.
Larson RL. 1988. The antioxidants of higher plants. Phytochemistry 4: 969-978.
Lasekan JB, Gueth L, Khan S. 1995. Influence of dietary golden pea proteins versus casein on plasma and hepatic lipids in rats. Nutrition Research 15(1): 71-84.
Lasztity R, Hidvegi M, Bata A. 1998. Saponins in food. Food Reviews International 14(4): 371-390.
Lau BH, Yamasaki T, Gridley DS. 1991. Garlic compounds modulate macrophage and T-lymphocyte functions. Molecular Biotherapy 3(2): 103-107.

Laughton JM, Evans PJ, Moroney MA, Hoult JRS. 1991. Inhibition of mammalian lipoxygenase and cyclo-oxygenase by flavonoid and phenolic dietary additives. Biochemical Pharmacology 18: 1673-1681.

Laurin J. 1931. Hypoglycaemic action of the bulbs of *Allium cepa* Linn. Comptes Rendus de l'Académie des Sciences de Paris 192: 1289.

Law LK, Tam PP, Yeung HW. 1983. Effects of alpha-trichosanthin and alpha-momorcharin on the development of peri-implantation mouse embryos. Journal of Reproduction and Fertility 69: 597-604.

Leatherdale BA, Panesar RK, Singh G, Atkins TW, Bailey CJ, et al. 1981. Improvement in glucose tolerance due to *Momordica charantia* (karela). Brazilian Medical Journal (Clinical Research Edition) 282: 1823-1824.

Lee KH, Choi HR, Kim CH. 2005. Anti-angiogenic effect of the seed extract of *Benincasa hispida* Cogniaux. Journal of Ethnopharmacology 97(3): 509-513.

Lee YM, Jeong HJ, Na HJ, Kim DK, Chae HJ, et al. 2001. Inhibition of immunologic and nonimmunologic stimulation-mediated anaphylactic shocks by water extract of white eggplant (*Solanum melongena*). Pharmacology Research 43(4): 405-409.

Lee SO, Lee IN. 2006. Induction of quinone reductase, the phase 2 anticarcinogenic marker enzyme, in Hepa1c1c7 cells by radish sprouts, *Raphanus sativus* L. Journal of Food Science 71(2): S144-S148.

Lee SC, Tsai CC, Chen JC, Lin CC, Hu ML, et al. 2002. The evaluation of reno- and hepatoprotective effects of huai-shan-yao (Rhizome Dioscoreae). American Journal of Clinical Medicine 30(4): 609-616.

Leighton T, Ginther C, Fluss L, Harter WK, Cansado J, et al. 1992. Quercetin and its glycosides in *Allium* vegetables. In: Ho C-T, Lee CY, Huang M-T, editors. Phenolic compounds in food and their effects on health II. American Chemical Society, Washington, DC. p. 220-238.

Le-Marchand L, Murphy SP, Hankin JH, Wilkens, LR, Kolonel LN. 2000. Intake of flavonoids and lung cancer. Journal of the National Cancer Institute 92: 154-160.

Leung SO, Yeung HW, Leung KN. 1987. The immunosuppressive activities of two abortifacient proteins isolated from the seeds of bitter melon (*Momordica charantia*). Immunopharmacology 13: 159-171.

Levey GA. 1993. The new power foods. Parade Magazine. Washington Post. November 14. p. 5.

Licastro F, Franceschi C, Barbieri L, Stirpe F. 1980. Toxicity of *Momordica charantia* lectin and inhibitor for human normal and leukaemic lymphocytes. Virchows Archives of B Cell Pathology Including Molecular Pathology 33: 257-265.

Liu Y, Perera CO, Suresh V. 2007. Comparison of three chosen vegetables with others from South East Asia for their lutein and zeaxanthin content. Food Chemistry 101(4): 1550-1556.

Lotufo LVC, Khan MTH, Ather A, Wilke DV, Jimenez PC, et al. 2005. Studies of the anticancer potential of plants used in Bangladeshi folk medicine. Journal of Ethnopharmacology 99(1): 21-30.

Mabey R. 1988. The new age herbalist. Macmillan, New York. p. 80.

Machlin LJ. 1984. Vitamin E. In: Machlin LJ, editor. Handbook of vitamins: Nutritional, biochemical and clinical aspects. Marcel Dekker, New York. p. 99-145.

Madar Z, Abel R, Samish S, Arad J. 1988. Glucose-lowering effect of fenugreek in non-insulin dependent diabetics. European Journal of Clinical Nutrition 42: 51-54.

Madar Z, Thorne R. 1987. Dietary fiber. Progress in Food and Nutrition Science 11(2): 153-174.

Mansour EH, Khalil AH. 2000. Evaluation of antioxidant activity of some plant extracts and their application to ground beef patties. Food Chemistry 69(2): 135-141.

Marcus DL, Thomas C, Rodriguez C, Simberkoff K, Tasi JS, et al. 1998. Increased peroxideratoin and reduced antioxidant enzyme activity in Alzheimer's disease. Experimental Neurology 150: 40-44.

Martins JM, Riottot M, de Abreu MC, Lança MJ, Viegas-Crespo AM, et al. 2004. Dietary raw peas (*Pisum sativum* L.) reduce plasma total and LDL cholesterol and hepatic esterified cholesterol in intact and ileorectal anastomosed pigs fed cholesterol-rich diets. The American Society for Nutritional Sciences. Journal of Nutrition 134: 3305-3312.

Matsuda H, Li Y, Yoshikawa M. 1999. Roles of capsaicin-sensitive sensory nerves, endogenous nitric oxide, sulfhydryls, and prostaglandins in gastroprotection by momordin Ic, an oleanolic acid oligoglycoside, on ethanol-induced gastric mucosal lesions in rats. Life Sciences 65: PL27-PL32.

Mayer B, Kalus U, Grigorov A, Pindur G, Jung F, et al. 2001. Effects of an onion-olive oil maceration product containing essential ingredients of the Mediterranean diet on blood pressure and blood fluidity. Arzneimittelforschung 51(2): 104-111.

McAnuff MA, Harding WW, Omoruyi FO, Jacobs H, Morrison EY, et al. 2005. Hypoglycemic effects of steroidal sapogenins isolated from Jamaican bitter yam, *Dioscorea polygonoides*. Food and Chemical Toxicology 43(11): 1667-1672.

McCarthy TL, Kerry JP, Kerry JF, Lynch PB, Buckley DJ. 2001a. Evaluation of the antioxidant potential of natural food/plant extracts as compared with synthetic antioxidants and vitamin E in raw and cooked pork patties. Meat Science 58(1): 45-52.

McCarthy TL, Kerry JP, Kerry JF, Lynch PB, Buckley DJ. 2001b. Assessment of the antioxidant potential of natural food and plant extracts in fresh and previously frozen pork patties. Meat Science 57(2): 177-184.

Middleton EJR, Kandaswami C. 1992. Effect of flavonoids on immune and inflammatory cell function. Biochemical Pharmacology 43(6): 1167-1179.

Miean KH, Mohamed S. 2001. Flavonoid (myricetin, quercetin, kaempferol, luteolin, and apigenin) content of edible tropical plants. Journal of Agricultural and Food Chemistry 49(6): 3106-3112.

Miller TE, Wing JS, Huete AR. 1984. The agricultural potential of selected C_4 plants in arid environment. Journal of Arid Environment 7: 275-286.

Milner JA. 1996. Garlic: Its anticarcinogenic and antitumorigenic properties. Nutrition Reviews 54: S82-S86.

Milner JA, Schaffer EM. 1998. Garlic and associated allyl sulfur constituents depress chemical carcinogenesis. ACS Symposium Series 702 (Functional Foods for Disease Prevention II: Medicinal Plants and Other Foods): 144-152.

Mohan M, Kaul N, Punekar A, Girnar R, Junnare P, et al. 2005. Nootropic activity of *Moringa oleifera* leaves. Journal of Natural Remedies 5(1): 59-62.

Morbidoni L, Arterburn, JM, Young V, Mullins D, Mulrow C, et al. 2001. Garlic: Its history and adverse effects. Journal of Herb Pharmacotherapy 1(1): 63-83.

Moundipa FP, Beboy NS, Zelefack F, Ngouela S, Tsamo E, et al. 2005. Effects of *Basella alba* and *Hibiscus macranthus* extracts on testosterone production of adult rat and bull Leydig cells. Asian Journal of Andrology 7(4): 411-417.

Moundipa FP, Kamtchouing P, Koueta N, Tantchou J, Foyang NP, et al. 1999. Effects of aqueous extracts of *Hibiscus macranthus* and *Basella alba* in mature rat testis function. Journal of Ethnopharmacology 65(2): 133-139.

Moundipa FP, Ngouela S, Kamtchouing P, Tsamo E, Tchouanguep FM, et al. 2006. Effects of extracts from *Hibiscus macranthus* and *Basella alba* mixture on testosterone production in vitro in adult rat testes slices. Asian Journal of Andrology 8(1): 111-114.

Moushumi L, Sumati B. 1994. Studies on possible protective effect of plant derived phenols and the vitamin precursors beta-carotene and alpha tocopherol on 7, 12 dimethylbenz[*a*]anthracene induce tumor initiation events. Phytotherapy Research 8(4): 237-240.

Munoz V, Sauvain M, Bourdy G, Callapa J, Rojas I, et al. 2000. The search for natural bioactive compounds through a multidisciplinary approach in Bolivia. Part II. Antimalarial activity of some plants used by *Mosetene indians*. Journal of Ethnopharmacology 69: 139-155.

Murakami T, Emoto A, Matsuda H, Yoshikawa M. 2001. Medicinal foodstuffs. Part XXI. Structures of new cucurbitane-type triterpene glycosides, goyaglycosides-a, -b, -c, -d, -e, -f, -g, and -h, and new oleanane-type triterpene saponins, goyasaponins I, II, and III, from the fresh fruit of Japanese *Momordica charantia* L. Chemical and Pharmaceutical Bulletin (Tokyo) 49: 54-63.

Murray MT. 1995. The healing power of herbs. 2nd ed. Prima Publishing, Rocklin, CA. p. 121-131.

Nadkarni KM. 1976. Indian materia medica. Popular Prakashan, Bombay.

Nagourney RA. 1998. Garlic: Medicinal food or nutritious medicine? Journal of Medicinal Food 1(1): 13-28.

Nair AGR, Subramanian SS. 1962. Pigments of the flowers of *Moringa pterygosperma*. Current Science 31: 155-156.

National Academies. 2001. Dietary intake recommendations for individuals, vitamins. Food and Nutrition Board, Institute of Medicine, National Academies. http://www.iom.edu/Object.File/Master/21/372/0.pdf. Accessed July 10, 2006.

National Research Council. 1989. Recommended dietary allowances. 10th ed. National Academy Press, Washington DC.

Nedic O, Nikolic JA, Hajdukovic-Dragojlovic L, Todorovic V, Masnikosa R. 2000. Alterations of IGF-binding proteins in patients with alcoholic liver cirrhosis. Alcohol 21(3): 223-229.

Nepal Department of Medicinal Plants. 1993. Medicinal plants of Nepal. Department of Medicinal Plants, Nepal.

Ng TB, Chan WY, Yeung HW. 1992. Proteins with abortifacient, ribosome inactivating, immunomodulatory, antitumor and anti-AIDS activities from Cucurbitaceae plants. General Pharmacology 23: 579-590.

Ng TB, Liu WK, Sze SF, Yeung HW. 1994. Action of alpha-momorcharin, a ribosome inactivating protein, on cultured tumor cell lines. General Pharmacology 25: 75-77.

Noda Y, Kaneyuki T, Igarashi K, Mori A, Packer L. 1998. Antioxidant activity of nasunin, an anthocyanin in eggplant. Research Communications in Molecular Pathology and Pharmacology 102: 175-187.

Noda Y, Kaneyuki T, Igarashi K, Mori A, Packer L. 2000. Antioxidant activity of nasunin, an anthocyanin in eggplant. Toxicology 148: 119-123.

Noguchi R, Yasui Y, Suzuki R, Hosokawa M, Fukunaga K, et al. 2001. Dietary effects of bitter gourd oil on blood and liver lipids of rats. Archives of Biochemistry and Biophysics 396: 207-212.

Odetola AA, Iranloye YO, Akinloye O. 2004. Hypolipidaemic potentials of *Solanum melongena* and *Solanum gilo* on hypercholesterolemic rabbits. Pakistan Journal of Nutrition 3(3): 180-187.

Ody P. 1993. The complete medicinal herbal. Dorling Kindersley, New York. p. 33.

Olmedilla B, Granado F, Blanco I, Vaquero M. 2003. Lutein, but not α-tocopherol, supplementation improves visual function in patients with age-related cataracts, a 2-y double blind placebo-controlled pilot study. Nutrition 19: 21-24.

Omara-Alwala TR, Mebrahtu T, Prior DE, Ezekwe MO. 1991. Omega-three fatty acids in purslane (*Portulaca oleracea*) tissues. Journal of the American Oil Chemists' Society 68: 198-199.

Omonigho SE. 1988. Deterioration of pounded yam and its shelf life extension (p. 162). MS Thesis. University of Benin, Benin City, Nigeria.

Omonigho SE, Ikenebomeh MJ. 2000. Effect of temperature treatment on the chemical composition of pounded white yam during storage. Food Chemistry 71: 215-220.

Orekhov AN, Grunwald J. 1997. Effects of garlic on atherosclerosis. Nutrition 13(7-8): 656-663.

Ovesna Z, Vachalkova A, Horvathova K. 2004. Taraxasterol and beta-sitosterol: New naturally compounds with chemoprotective/chemopreventive effects. Neoplasma 51(6): 407-414.

Pal SK, Mukherjee PK, Saha BP. 1995. Studies on the antiulcer activity of *Moringa oleifera* leaf extract on gastric ulcer models in rats. Phytotherapy Research 9: 463-465.

Palaniswamy UR. 1998. Enhancement of naturally occurring chemopreventive compounds in salad greens through environmental manipulation during crop growth. Dissertation. University of Connecticut, Storrs.

Pari L, Uma A. 2003. Protective effect of *Sesbania grandiflora* against erythromycin estolate-induced hepatotoxicity. Therapie 58(5): 439-443.

Parkash A, Ng TB, Tso WW. 2002. Purification and characterization of charantin, a napin-like ribosome-inactivating peptide from bitter gourd (*Momordica charantia*) seeds. Journal of Peptide Research 59: 197-202.

Patterson RE, White E, Kristal AR, Neuhouser ML, Potter JD. 1997. Vitamin supplements and cancer risk: The epidemiologic evidence. Cancer Causes and Control 8: 786-802.

Pedraza-Chaverri J, Tapia E, Medina-Campos ON, de los Angeles GM, Franco M. 1998. Garlic prevents hypertension induced by chronic inhibition of nitric oxide synthesis. Life Sciences 62(6): PL71-77.

Petit PR, Sauviaire YD, Buys DMH, Leconte OM, Baissac YG, et al. 1995. Steroid saponins from fenugreek seeds: Extraction, purification and pharmacological investigation on feeding behavior and plasma cholesterol. Steroids 10: 674-680.

Phelps S, Harris WS. 1993. Garlic supplementation and lipoprotein oxidation susceptibility. Lipids 28(5): 475-477.

Pineau N, Aucouturier P, Preud'Homme JL, Hagiwara K, Kobayashi K. 1991. Structural and functional variability of jacalin. Molecular Immunology 28(1-2): 185-187.

Pineau N, Pousset JL, Preud'Homme JL, Aucouturier P. 1990. Structural and functional similarities of breadfruit seed lectin and jacalin. Molecular Immunology 27(3): 237-240.

Pinto JT, Qiao C, Xing J, Rivlin RS, Protomastro ML, et al. 1997. Effects of garlic thioallyl derivatives on growth, glutathione concentration, and polyamine formation of human prostate carcinoma cells in culture. American Journal of Clinical Nutrition 66(2): 398-405.

Platel K, Shurpalekar KS, Srinivasan K. 1993. Influence of bitter gourd (*Momordica charantia*) on growth and blood constituents in albino rats. Nahrung 37: 156-160.

Platel K, Srinivasan K. 1997. Plant foods in the management of diabetes mellitus: Vegetables as potential hypoglycemic agents. Nahrung 41(2): 68-74.

Potter JD, Steinmetz K. 1996. Vegetables, fruits, and phytoestrogens as preventive agents. IARC Scientific Publications 139: 61-90.

Prasad G, Sahay SS, Masood A. 1994. Inhibition in aflatoxin biosynthesis by the extract of *Amorphophallus campanulatus* (OL) and calcium oxalate. Letters in Applied Microbiology 18(4): 203-205.

Prathibha S, Nambisan B, Leelamma S. 1995. Enzyme inhibitors in tuber crops and their thermal stability. Plant Foods for Human Nutrition 48(3): 247-257.

Produce for Better Health Foundation. 2007. http://www.5aday.com/html/consumers/colorway.php. Accessed June 20, 2007.

Proeston SR. 1995. Aibika/Bele—*Abelmoschus manihot* (L.) Medik. CAB Abstracts, CAB International (1998). p. 97.

Puel C, Mathey J, Kati-Coulibaly S, Davicco MJ, Lebecque P, et al. 2005. Preventive effect of *Abelmoschus manihot* (L.) Medik. on bone loss in the ovariectomised rats. Journal of Ethnopharmacology 99(1): 55-60.

Punna R, Rao PU. 2004. Effect of maturity and processing on total, insoluble and soluble dietary fiber contents of Indian green leafy vegetables. International Journal of Food Science and Nutrition 55(7): 561-567.

Raghuram TC, Sharma RD, Sivakumar B, Sahay BK. 1994. Effect of fenugreek seeds on intravenous glucose disposition in non-insulin dependent diabetic patients. Phytotherapy Research 8: 83-86.

Rahmat A, Kumar V, Fong LM, Endrini S, Sani HA. 2004. Determination of total antioxidant activity in three types of local vegetables shoots and the cytotoxic effect of their ethanolic extracts against different cancer cell lines. Asia Pacific Journal of Clinical Nutrition 13(3): 308-311.

Rajyalakshmi P, Venkatalaxmi K, Venkatalakshmamma K, Jyothsna Y, Devi KB, et al. 2001. Total carotenoid and beta-carotene contents of forest green leafy vegetables consumed by tribals of south India. Plant Foods for Human Nutrition 56(3): 225-238.

Ramachandran C, Peter KV, Gopalakrishnan PK. 1980a. Drumstick (*Moringa oleifera*): A multipurpose Indian vegetable. Economic Botany 34(3): 276-283.

Ramachandran C, Peter KV, Gopalakrishnan PK. 1980b. Chekurmanis a multi-vit leafy vegetable. Indian Horticulture 25(1): 17.

Raman A, Lau C. 1996. Anti-diabetic properties and phytochemistry of *Momordica charantia* L. (Cucurbitaceae). Phytomedicine 2: 349-362.

Ramdath DD, Isaacs RLC, Teelucksingh S, Wolever TMS. 2004. Glycaemic index of selected staples commonly eaten in the Caribbean and the effects of boiling v. crushing. British Journal of Nutrition 91(6): 971-977.

Rangaswamy L, Marisiddaiah R, Thirumalai PK, Vallikannan B. 2005. Determination of major carotenoids in a few Indian leafy vegetables by high-performance liquid chromatography. Journal of Agricultural and Food Chemistry 53(8): 2838-2842.

Rao GMM, Rao CV, Sudhakara M, Pandey MM, Rawat AKS, et al. 2004. Anti-inflammatory and antinociceptive activities of "*Coccinia indica* W. & A." fruit juice powder in animals. Natural Product Sciences 10(1): 20-23.

Rao NV, Prasad MVV, Raju YP. 2005. Hepatoprotective activity of *Coccinia grandis* (Linn.) fruit extracts on carbon tetrachloride induced liver damage in rats. Asian Journal of Chemistry 17(3): 1780-1784.

Rao PU, Sesikeran B, Rao PS, Naidu AN, Rao VV, et al. 1996. Short term nutritional and safety evaluation of fenugreek. Nutrition Research 16(9): 1495-1505.

Rasfon, K. 1991. The anticestodal activity of preparations made from the breadfruit. Meditsinskaia Parazitologiia i Parazitarnye Bolezni 5: 49-52.

Rashidkhani B, Lindblad P, Wolk A. 2005. Fruits, vegetables and risk of renal cell carcinoma: A prospective study of Swedish women. International Journal of Cancer 113(3): 451-455.

Rathi SS, Grover JK, Vats V, Biswas NR. 2002. Prevention of experimental diabetic cataract by Indian Ayurvedic plant extracts. Phytotherapy Research 16: 774-777.

Ravikumar P, Anuradha CV. 1999. Effect of fenugreek seeds on blood lipid peroxidation and antioxidants in diabetic rats. Phytotherapy Research 13(3): 197-201.

Remani P, Bhattathiri VN, Bindu L, Chandralekha B, Vijayakumar T, et al. 1997. Correlation of lectin binding with lymph node metastasis in oral cancers. Oral Oncology 33(1): 19-22.

Resmi DS, Celine VA, Rajamony L. 2005. Variability among drumstick (*Moringa oleifera* Lam.) accessions from central and southern Kerala. Journal of Tropical Agriculture 43(1-2): 83-85.

Rochfort SJ, Imsic M, Jones R, Trenerry VC, Tomkins B. 2006. Characterization of flavonol conjugates in immature leaves of pak choi [*Brassica rapa* L. Ssp. chinensis L. (Hanelt.)] by HPLC-DAD and LC-MS/MS. Journal of Agricultural and Food Chemistry 54(13): 4855-4860.

Roman-Ramos R, Flores-Saenz JL, Alarcon-Aguilar FJ. 1995. Anti-hyperglycemic effect of some edible plants. Journal of Ethnopharmacology 48(1): 25-32.

Sahu BR, Chakrabarty A. 1994. Screening of antibacterial activity of various extractives of *Cassia tora* and *Alternanthera sessilis*. Asian Journal of Chemistry 6(3): 687-689.

Sai KS, Srividya N. 2002. Blood glucose lowering effect of the leaves of *Tinospora cordifolia* and *Sauropus androgynus* in diabetic subjects. Journal of Natural Remedies 2(1): 28-32.

Sampson L, Rimm E, Hollman PCH, de Vries JHM, Katan MB. 2002. Flavonol and flavone intakes in US health professionals. Journal of American Dietetic Association 102(10): 1414-1420.

Sánchez-Machado DI, López-Cervantes J, Ríos Vázquez NJ. 2006. High-performance liquid chromatography method to measure α- and γ-tocopherol in leaves, flowers and fresh beans from *Moringa oleifera*. Journal of Chromatography 1105 (1-2) (Special Issue): 111-114.

Sankaranarayanan R, Sekar K, Banerjee R, Sharma V, Surolia A, et al. 1996. A novel mode of carbohydrate recognition in jacalin, a Moraceae plant lectin with a beta-prism fold. Nature Structural Biology 3(7): 596-603.

Satyawati GV, Gupta AK, Tandon N. 1987. Medicinal plants of India. Indian Council of Medical Research, New Delhi. p. 262.

Schreiber CA, Wan L, Sun Y, Lu L, Krey LC, et al. 1999. The antiviral agents, MAP30 and GAP31, are not toxic to human spermatozoa and may be useful in preventing the sexual transmission of human immunodeficiency virus type 1. Fertility and Sterility 72: 686-690.

Şehirali S. 1988. Yemeklik dane baklagiller [Pulses]. Ankara, Ankara University Press, Turkey.

Semba RD, Dagnelie G. 2003. Are lutein and zeaxanthin conditionally essential nutrients for eye health? Medical Hypotheses 61(4): 465-472.

Sesso HD, Gaziano JM, Liu S, Buring JE. 2003. Flavonoid intake and the risk of cardiovascular disease in women. American Journal of Clinical Nutrition 77(6): 1400-1408.

Setchell KD, Lawson AM, Borriello SP, Harkness R, Gordon H, et al. 1981. Lignan formation in man: Microbial involvement and possible roles in relation to cancer. Lancet 318(8236): 4-7.

Shahidi F, and Naczk M., 1995. Food phenolics: Sources, chemistry, effects and applications. Technomic Publishing Co., Lancaster, PA.

Sharma RD, Raghuram TC, Dayasagar RV. 1991. Hypolipidaemic effect of fenugreek seeds. A clinical study. Phytotherapy Research 5: 145-147.

Sharma RD, Raghuram TC, Rao NS. 1990. Effect of fenugreek seeds on blood glucose and serum lipids in type I diabetes. European Journal of Clinical Nutrition 44(4): 301-306.

Sharma RD, Sarkar A, Hazra DK. 1996. Hypolipidaemic effect of fenugreek seeds: A chronic study in non-insulin dependent diabetic patient. Phytotherapy Research 10: 332-334.

Sheela CG, Kumud, K, Augusti KT. 1995. Anti-diabetic effects of onion and garlic sulfoxide amino acids in rats. Planta Medica 61: 356-367.

Shibib BA, Khan LA, Rahman R. 1993. Activity of *Coccinia indica* and *Momordica charantia* in diabetic rats: Depression of the hepatic gluconeogenic enzymes glucose-6-phosphatase and fructose-1,6-bisphosphatase and elevation of both liver and red-cell shunt enzyme glucose-6-phosphate dehydrogenase. Biochemistry Journal 292: 267-270.

Shilpi JA, Ray PK, Sarder MM, Uddin SJ. 2005. Analgesic activity of *Amorphophallus campanulatus* tuber. Fitoterapia 76(3-4): 367-369.

Shrikhande AJ, Franch FJ. 1974. Effect of flavonols on ascorbic acid and anthocyanin stability in model systems. Journal of Food Science 39(5): 904-906.

Siddhuraju P, Becker K. 2003. Antioxidant properties of various solvent extracts of total phenolic constituents from three different agroclimatic origins of drumstick tree (*Moringa oleifera* Lam.) leaves. Journal of Agricultural and Food Chemistry 51(8): 2144-2155.

Siddiq A, Anwar F, Manzoor M, Fatima A. 2005. Antioxidant activity of different solvent extracts of Moringa oleifera leaves under accelerated storage of sunflower oil. Asian Journal of Plant Sciences 4(6): 630-635.

Silagy CA, Neil HAA. 1994. Meta-analysis of the effect of garlic on blood pressure. Journal of Hypertension 12(4): 463-468.

Simopoulos AP. 1999. Essential fatty acids in health and chronic disease. American Journal of Clinical Nutrition 70(3 suppl.): 560S-569S.

Simopoulos AP, Norman HA, Gillapsy JE, Duke JA. 1992. Common purslane: A source of omega-3 fatty acids and antioxidants. Journal of the American College of Nutrition 11: 374-382.

Simopoulos AP, Salem Jr. N. 1986. Purslane: A terrestrial source of omega-3 fatty acids. New England Journal of Medicine 315: 833.

Singh N, Tyagi SD, Agarwal SC. 1989. Effects of long term feeding of acetone extract of *Momordica charantia* (whole fruit powder) on alloxan diabetic albino rats. Indian Journal of Physiology and Pharmacology 33: 97-100.

Smith AC. 1981. Flora *Vitiensis nova:* A new flora of Fiji. Lawai, Kauai, Hawaii. National Tropical Botanical Garden 2: 291-292.

Smith TJ, Yang CS. 1994. Effect of food phytochemical on metabolism and tumorigenesis. In: Huang MT, editor. Food phytochemicals for cancer prevention I. American Chemical Society, Washington, DC. p. 48.

Solis CS. 1969. Antibacterial and antibiotic properties of the Leguminosae. Acta Manilana Series A(4): 52-109.

Sommerburg OG, Siems WG, Hurst JS, Lewis JW, Kliger DS, et al. 1999. Lutein and zeaxanthin are associated with photoreceptors in the human retina. Current Eye Research 19: 491-495.

Spiliotis V, Lalas S, Gergis V, Dourtoglou V. 1998. Comparison of antimicrobial activity of seeds of different *Moringa oleifera* varieties. Pharmaceutical and Pharmacological Letters 8(1): 39-40.

Spreafico F, Malfiore C, Moras ML, Marmonti L, Filippeschi S, et al. 1983. The immunomodulatory activity of the plant proteins *Momordica charantia* inhibitor and pokeweed antiviral protein. International Journal of Immunopharmacology 5: 335-343.

Sreedevi I, Chaturvedi A. 1993. Effect of vegetable fiber on postprandial glycemia. Plant Foods for Human Nutrition 44(1): 71-78.

Sridhar R, Lakshminarayana G. 1993. Lipid classes, fatty acids, and tocopherols of leaves of six edible plant species. Journal of Agricultural and Food Chemistry 41(1): 61-63.

Stark A, Madar Z. 1993. The effect of an ethanol extract derived from fenugreek (*Trigonella foenum-graecum*) on bile acid absorption and cholesterol levels in rats. British Journal of Nutrition 69(1): 277-287.

Steiner M, Khan AH, Holbert D, Lin RI. 1996. A double-blind crossover study in moderately hypercholesterolemic men that compared the effect of aged garlic extract and placebo administration on blood lipids. American Journal of Clinical Nutrition 64(6): 866-870.

Stevinson C, Pittler MH, Ernst E. 2000. Garlic for treating hypercholesterolemia: A meta-analysis of randomized clinical trials. Annals of Internal Medicine 133(6): 420-429.

Stewart JR, Artime MC, O'Brian CA. 2003. Resveratrol: A candidate nutritional substance for prostate cancer prevention. Journal of Nutrition 133(7 suppl.): 2440S-2443S.

Sudhakar M, Rao AL. 2005. Antimicrobial activity of *Tinospora cordifolia, talinum cuneiformis* and *coccinia indica*. International Journal of Chemical Science 3(3): 523-528.

Sudheesh S, Presannakumar G, Vijayakumar S, Vijayalakshmi RN. 1997. Hypolipidemic effect of flavonoids from *Solanum melongena*. Plant Foods for Human Nutrition 51: 321-330.

Sudheesh S, Sandhya C, Koshy SA, Vijayalakshmi NR. 1999. Antioxidant activity of flavonoids from *Solanum melongena*. Phytotherapy Research 13: 393-396.

Sujathan K, Kannan S, Remani P, Pillai KR, Chandralekha B, et al. 1996. Differential expression of jackfruit-lectin-specific glycoconjugates in metastatic adenocarcinoma and reactive mesothelial cells: A diagnostic aid in effusion cytology. Journal of Cancer Research and Clinical Oncology 122(7): 433-436.

Sur P, Das M, Gomes A, Vedasiromoni JR, Sahu NP, et al. 2001. *Trigonella foenum graecum* (fenugreek) seed extract as an antineoplastic agent. Phytotherapy Research 15(3): 257-259.

Suresh B, Srinivasan K. 1997. Influence of dietary capsaicin and onion on the metabolic abnormalities associated with streptozotocin induced diabetes mellitus. Molecular and Cellular Biochemistry 175(1-2): 49-57.

Suresh B, Srinivasan K. 1999. Renal lesions in streptozotocin-induced diabetic rats maintained on onion and capsaicin containing diets. Journal of Nutritional Biochemistry 10(8): 477-483.

Świderski F. (red.). 2000. Żywność wygodna izywność funkcjonalna. WNT, Warszawa (in Polish).
Tahiliani P, Kar A. 2000. Role of *Moringa oleifera* leaf extract in the regulation of thyroid hormone status in adult male and female rats. Pharmacology Research 41(3): 319-323.
Takemoto DJ, Dunford C, Vaughn D, Kramer KJ, Smith A, et al. 1982. Guanylate cyclase activity in human leukemic and normal lymphocytes. Enzyme inhibition and cytotoxicity of plant extracts. Enzyme 27: 179-188.
Tan W, Lin DX, Xiao Y, Kadlubar F, Chen JS. 1999. Chemoprevention of 2-amino-1-methyl-6-phenyli-midazo 4,5-b pyridine-induced carcinogen-DNA adducts by Chinese cabbage in rats. World Journal of Gastroenterology 5(2): 138-142.
Tayyaba Z, Hasnain SM, Hasan SK. 2001. Evaluation of the oral hypoglycemic effects of *Trigonella foenum graecum* L. (Methi) in normal mice. Journal of Ethnopharmacology 75: 191-195.
Terry P, Wolk A, Persson I, Magnusson C. 2001. Brassica vegetables and breast cancer risk. Journal of the American Medical Association 285(23): 2975-2977.
Traber MG, Burton GW, Ingold KU, Kayden HJ. 1990. RRR- and SRR-alpha-tocopherols are secreted without discrimination in human chylomicrons, but RRR-alpha-tocopherol is preferentially secreted in very low-density lipoproteins. Journal of Lipid Research 31: 675-685.
Tsushida, T, Suzuki, M. 1995. Isolation of flavonoid-glycosides in onion and identification by chemical synthesis of the glycosides. Nipp Shok Kaga Koga Kaishi 42(2): 100-108.
Undie AS, Akubue PI. 1986. Pharmacological evaluation of *Dioscorea dumetorum* tuber used in traditional antidiabetic therapy. Jounal of Ethnopharmacology 15: 133-144.
Vajifdar BU, Goyal VS, Lokhandwala YY, Mhamunkar SR, Mahadik SP, et al. 2000. Is dietary fiber beneficial in chronic ischemic heart disease? Journal of Association of the Physicians of India 48(9): 871-876.
Verreault R, Chu J, Mandelson M, Shy KA. 1989. Case-control study of diet and invasive cervical cancer. International Journal of Cancer 43: 1050-1054.
Vesely DL, Graves WR, Lo TM, Fletcher MA, Levey GS. 1977. Isolation of a guanylate cyclase inhibitor from the balsam pear (*Momordica charantia* abreviata). Biochemical and Biophysical Research Communications 77: 1294-1299.
Vijayan KK, Remani P, Beevi VM, Ankathil R, Vijayakumar T, et al. 1987. Tissue binding patterns of lectins in premalignant and malignant lesions of the oral cavity. Journal of Experimental Pathology 3(3): 295-304.
Vikrant V, Grover JK, Tandon N, Rathi SS, Gupta N. 2001. Treatment with extracts of *Momordica charantia* and *Eugenia jambolana* prevents hyperglycemia and hyperinsulinemia in fructose fed rats. Journal of Ethnopharmacology 76: 139-143.
Vinson JA, Dabbagh YA, Serry MM, Jang J. 1995. Plant flavonoids, especially tea flavonols, are powerful antioxidants using an *in vitro* oxidation model for heart disease. Journal of Agricultural and Food Chemistry 43: 2800-2804.
Vohora SB, Kumar I, Khan MS. 1984. Effect of alkaloids of *Solanum melongena* on the central nervous system. Journal of Ethnopharmacology 11(3): 331-336.

Wanasundera JPD, Ravindran G. 1994. Nutrition assessment of yam (*Dioscorea alata*) tubers. Plant Foods for Human Nutrition 46: 33-39.

Wang HX, Ng TB. 2001. Studies on the anti-mitogenic, anti-phage and hypotensive effects of several ribosome inactivating proteins. Comparative Biochemistry and Physiology C-Pharmacology Toxicology 128: 359-366.

Watson RR, Leonard TK. 1986. Selenium and vitamins A, E and C: Nutrients with cancer prevention properties. Journal of American Dietetics Association 86: 505-510.

Weber P, Bendich A, Machlin L. 1997. Vitamin E and human health: Rationale for determining recommended intake levels. Nutrition 13: 450-460.

Welihinda J, Karunanayake EH, Sheriff MH, Jayasinghe KS. 1986. Effect of *Momordica charantia* on the glucose tolerance in maturity onset diabetes. Journal of Ethnopharmacology 17: 277-282.

Wenzel GE, Fontana JD, Correa JBC. 1990. The viscous mucilage from the weed *Portulaca oleracea* L. Applied Biochemistry and Biotechnology 24-25: 341-353.

Wettasinghe M, Bolling B, Plhak L, Parkin K. 2002. Screening for phase II enzyme-inducing and antioxidant activities of common vegetables. Journal of Food Science 67: 2583-2588.

Wills RBH, Rangga A. 1996. Determination of carotenoids in Chinese vegetables. Food Chemistry 56(4): 451-455.

World Cancer Research Fund. 1997. Food, nutrition and the prevention of cancer: A global perspective. American Institute for Cancer Research, Washington, DC.

Wu WH, Liu LY, Chung CJ, Jou HJ, Wang TA. 2005. Estrogenic effect of yam ingestion in healthy postmenopausal women. Journal of the American College of Nutrition 24(4): 235-243.

Xing N, Chen Y, Mitchell SH, Young CYF. 2001. Quercetin inhibits the expression and function of the androgen receptor in LNCaP prostate cancer cells. Carcinogenesis 22: 409-414.

Xu Y, Liu H, Dao X. 2003. Nutritional contents of *Coccinia grandis* and its evaluation as a wild vegetable. Yunnan Zhiwu Yanjiu 25(6): 680-686.

Yadav RN, Syeda Y. 1994. Chemical examination of the seeds of *Trichosanthes anguina* Linn. Asian Journal of Chemistry 6(1): 182-184.

Yeh YY, Liu L. 2001. Cholesterol-lowering effect of garlic extracts and organosulfur compounds: Human and animal studies. Journal of Nutrition 131(3 suppl.): 989S-993S.

Yen GC, Chen HY, Peng HH. 2001. Evaluation of the cytotoxicity, mutagenicity and antimutagenicity of emerging edible plants. Food Chemistry and Toxicology 39(11): 1045-1053.

Yen KY. 1992. The illustrated Chinese materia medica. SMC Publishing, Taipei. p. 84.

Yesilada E, Gurbuz I, Shibata H. 1999. Screening of Turkish anti-ulcerogenic folk remedies for anti-*Helicobacter pylori* activity. Journal of Ethnopharmacology 66: 289-293.

Yeum KJ, Taylor A, Tang G, Russell RM. 1995. Measurement of carotenoids, retinoids and tocopherols in human lens. Investigative Ophthalmology and Visual Science 36: 2756-2761.

Yoshida M, Sakai T, Hosokawa N, Marui N, Matsumoto K, et al. 1990. The effect of quercetin on cell cycle progression and growth of human gastric cancer cells. FEBS Letters 260(1): 10-13.

Yoshizumi S, Murakami T, Kadoya M, Matsuda H, Yamahara J, et al. 1998. Medicinal foodstuffs. XI. Histamine release inhibitors from wax gourd, the fruits of *Benincasa hispida* Cogn. Yakugaku Zasshi 118: 188-192.

Young G, Lawrence R, Schreuder M. (2005). Discovery of the ultimate superfood: How the Ningxia Wolfberry and four other foods help combat heart disease, cancer, chronic fatigue, depression, diabetes and more. Essential Science Publishing, Orem, Utah.

Yuan YR, He YN, Xiong JP, Xia ZX. 1999. Three-dimensional structure of beta-momorcharin at 2.55 A resolution. Acta Crystallographica Section D-Biological Crystallography 55: 1144-1151.

Yusuf M, Chowdhury JU, Yahab MA, Begum J. 1994. Medicinal plants of Bangladesh. BCSIR Laboratories, Bangladesh. p. 20.

Zhang Y, Talalay P. 1994. Anticarcinogenic activities of organic isothiocyanates: Chemistry and mechanisms. Cancer Research 54(7 suppl.): 1976S-1981S.

Zheng YT, Ben KL, Jin SW. 1999. Alpha-momorcharin inhibits HIV-1 replication in acutely but not chronically infected T-lymphocytes. Zhongguo Yao Li Xue Bao 20: 239-243.

Zia T, Hasnain SN, Hasan SK. 2001. Evaluation of the oral hypoglycaemic effect of *Trigonella foenum-graecum* L. (methi) in normal mice. Journal of Ethnopharmacology 75(2-3): 191-195.

Zielinski H, Frias J, Piskula MK, Kozlowska H, Vidal-Valverde C. 2005. Vitamin B1 and B2, dietary fiber and minerals content of Cruciferae sprouts. European Food Research and Technology 221(1-2): 78-83.

Chapter 3

Fruits

The botanical term *fruit* refers to both fruits and vegetables. Vegetables are immature fruits consumed raw or cooked, while fruits are consumed in their ripe form and do not need to be cooked before consumption. Traditionally valued for their tasty pulp, or meat, and as excellent sources of ascorbic acid and other vitamins and essential minerals, fruits are now valued for their peels and pigment content as well. Since pigments and other plant secondary metabolites, which are essentially plant defense compounds, are more concentrated in fruit peel than pulp, the peel has a functional value in human health. Anthocyanins, a major group of plant pigments, develop as the fruits ripen, and these flavonoids are often responsible for the orange, red, and blue colors in fruits. Fruits and their anthocyanins have come to be known for their functional benefits, including their strong antioxidant activity in metabolic reactions (Einbond et al. 2004), inhibition of cancer cell growth (Kamei et al. 1993), chemotherapeutic properties for numerous diseases (Bomser et al. 1996), and effect on oxidative stress, the aging process, and cancers (Rice-Evans 1999). Antioxidant properties of anthocyanins are due to their high reactivity as hydrogen or electron donors, the ability of their polyphenol-derived radicals to stabilize and delocalize the unpaired electrons, and their ability to chelate transition metal ions (Rice-Evans et al. 1997).

POME FRUITS

*Apple (***Malus domestica**; *Rosaceae)*

Apples are one of the most ancient fruit trees; they are part of the creation story of Adam and Eve. They have long been associated with

doi:10.1300/5504_03

human health, as in the adage "An apple a day keeps the doctor away," and recent studies have added extra weight to the old adage. They are small to medium-sized trees with a spreading canopy and elliptical, dark green leaves with serrate margins and light pubescence on their underside. The inflorescence is a cyme of four to six flowers, with the center flower opening first. The flowers are produced terminally from mixed buds (containing both leaves and flowers) on spurs. The fruits are characteristic and are called *pomes*. The fleshy edible portion is derived from the hypanthium, or floral cup, not the ovary. The locular fruit contains five seed cavities with generally two seeds each. The seeds are small and black.

There are five major groups of polyphenolic compounds in apples—hydroxycinnamic acids, flavan-3-ols, anthocyanidins, flavonols, and dihydrochalcones—whose amounts vary with the growth period, growing season, geographic location, and genetic variation (Tsao et al. 2003). The concentration and composition of phytochemicals in apples vary greatly between the apple peel and the apple flesh. Depending on the variety, compared with the pulp, apple peel contains two to six times more phenolic compounds and two to three times more flavonoids (Lata et al. 2005). The most predominant phenolic compounds in apple peel are phenolic acids (hydroxycinnamic acid derivatives such as chlorogenic acid or caffeic acid) and the flavonoids. Table 3.1 gives a dietary fiber profile and Table 3.2 a phenolic profile. Recent epidemiological and in vitro experimental studies have shown the linkage between apple and related products in the diet and many chronic human diseases caused by oxidative processes, especially those involving excess free radicals and reactive oxygen species,

TABLE 3.1. Dietary Fiber Content in Apple and Pear Peel and Pulp (g/kg FW)

Samples	Total	Soluble	Insoluble
Apple peel	28.9	10.9	18.0
Apple pulp	22.2	8.9	13.3
Pear peel	28.1	10.2	17.9
Pear pulp	21.3	8.4	12.9

Source: Leontowicz et al. 2003.

Note: FW = fresh weight.

TABLE 3.2. Content of Phenolic Acids in Apple and Pear Peel and Pulp (g/kg FW)

Samples	Ferulic Acid	*p*-Coumaric	Caffeic Acid
Apple peel	136.7	523.1	2,597
Apple pulp	119.8	367.9	1,993
Pear peel	49.1	150.9	877
Pear pulp	21.3	111.7	701

Source: Leontowicz et al. 2003.

Note: FW = fresh weight.

such as lowered risk of cancers (particularly prostate, liver, colon, and lung; McCann et al. 2007; Willett 1995; Xing et al. 2001) and cardiovascular disease (Hyson et al. 2000; Knekt et al. 1996). Consumption of apples is also suggested as an effective strategy for breast cancer protection (Liu et al. 2005).

In 2005 global apple production was 63,488,907 Mt and the area harvested was 5,218,126 ha (FAOSTAT 2006).

*Pear (*Pyrus communis, P. pyrifolia; *Rosaceae)*

Pyrus species are thought to have originated in the mountainous area of western China. *Pyrus* includes the Oriental pear and the Occidental pear. Oriental pears are distributed mainly around eastern Asia, including China, the Korean peninsula, and Japan; major cultivated species include *P. bretschnrideri, P. pyrifolia, P. ussuriensis,* and *P. sinkiangensis*. In China the pear has been considered not only a fruit but also an herbal medicine with antitussive, anti-inflammatory, and diuretic effects and has been used for more than 1,500 years (Li 1982). The plant is a medium-sized tree whose leaves are elliptic or ovate with acute tips and finely serrate or entire margins. Inflorescence is a corymbose, containing five to seven white flowers borne from terminal mixed buds on short spurs that occur on two-year and older wood. The fruit is a pome with five central seed cavities, usually bearing two seeds each, as do apples. Pears are a rich source of phenols and antioxidant vitamins, and they exhibit antioxidant activity and are antiproliferative against cancer cell lines (Sun et al. 2002). Cui et al. (2005) reported arbutin and chlorogenic acid as the major phenolic constituents of *Pyrus* species. See Tables 3.1 and 3.2 for dietary fiber and phenolic content profiles.

In 2005 global pear production was 19,553,799 Mt and the area harvested was 1,741,188 ha (FAOSTAT 2006).

DRUPES, OR STONE FRUITS

*Peach (***Prunus persica***; Rosaceae)*

The peach is native to China. The smooth-skinned nectarine is a cultivar of the peach. The peach is a small deciduous tree growing to 5 to 10 m. The leaves are lanceolate; flowers are solitary or paired, 2.5 to 3 cm in diameter, pink, and have five petals. The fruit is a drupe, with a single large seed encased in hard wood (called the stone or pit), and has yellow or whitish flesh, a delicate aroma, a velvety skin, and a tendency to bruise easily.

In traditional medicine, peach flowers are considered purgative and a decoction of the leaves is used as an anthelmintic, laxative, and sedative (Nadkarni 1976). Phytochemical studies on peach leaves revealed the presence of multiple chemicals, such as caffeic acid, chlorogenic acid, *p*-coumaric acid and ester, flavonol-glycosides, kaempferol, quercetin, quercetin-3-glycoside, quercetin-3-rhamnoside, quinic acid, tannin, ursolic acid, zeaxanthin (Duke 1992) and prussic acid (Nadkarni 1976). Pharmacological studies on peach leaves showed them to have multiple activities, including antimalarial (Misra et al. 1991), hepatoprotective (Lee et al. 1992), antiasthmatic (Ding 1983), and anticoagulant (Kosuge et al. 1985). Similarly, peach oil has been reported to have antifungal activity (Mishra and Dubey 1990). Peach fruit is a good source of fiber and anthocyanin pigments (Grigelmo-Miguel et al. 1999) and hence has great potential as an antioxidant.

*Apricot (***P. armeniaca** *L.; Rosaceae)*

The apricot is a moderate-sized tree, widely distributed in the northwestern Himalayas in India and many temperate and subtropical regions. It has been reported as an antipyretic, a refrigerant, and a thirst quencher and has been used to treat leprosy and leukoderma (Chopra et al. 1956; Kirtikar and Basu 1974). It is used in Unani medicine as an antidiarrheal, antipyretic, emetic, and anthelmintic; to treat liver diseases, piles, earache, and deafness; and its anti-inflammatory and analgesic activities have been reported (Kirtikar and Basu 1974;

Nagamoto et al. 1988). It is an excellent source of carotenoids, 1,512 to 16,500 μg/100 g edible portion (Ruiz et al. 2005).

*Cherry (*P. cerasus*; Rosaceae)*

The cherry is a tree with a broad crown or it is grown as a bush. The leaves are alternate, oval to obovate, with serrate margins; five-petal flowers are white and borne in clusters of three or five; and fruits are dark red to black. Antioxidant and anti-inflammatory activities of anthocyanins extracted from cherries have led researchers to study this fruit as a functional food (Wang et al. 1999). The antioxidants in sour cherries have potentials comparable to those found in some berry fruits (e.g., strawberry) and higher than those in apple and kiwi fruit (Blando et al. 2004). Seed extract showed cardioprotective activity in some studies (Bak et al. 2006).

OTHER FRUITS

*Mango (*Mangifera indica*; Anacardiaceae)*

Indigenous to India and Burma where it still occurs in the wild, the mango is one of the most celebrated of tropical fruits. It is one of the Dravidian Tamil culture's royal trinity of fruits (mango, banana, and the jackfruit). Mangoes have been cultivated for more than 4,000 years for their edible fruits, ornamental leaves, and shade. The mango is named the king of the fruits and called the apple of the tropics, ranking only second (after pineapple) among internationally traded tropical fruits in quantity and value. The interest in the mango as an edible fruit is so great that more than 1,000 varieties of it have been developed, of which at least 65 varieties are commercially cultivated on a large scale around the world.

India is a major exporter of mangoes, shipping almost two-thirds of its national production. It is a major fruit produced during the summer season in Taiwan. Other leading growers and exporters of fresh mangoes are the Philippines, Thailand, and Mexico. Mangoes are cultivated in other tropical countries, including Australia, Sri Lanka, Indonesia, Tanzania, Pakistan, Bangladesh, Brazil, Kenya, Egypt, and Sudan.

Dried mango flowers are used by Ayurvedic practitioners to treat diarrhea, chronic dysentery, and chronic urethritis. The bark is an astringent and used to treat rheumatism and diphtheria in India. The gum oozing from the trunk is collected and applied on cracked skin of the feet and on scabies. Mango kernel powder is used as a culinary spice and anthelmintic and as an astringent for diarrhea, hemorrhages, and bleeding hemorrhoids. Extracts of unripe fruits and of bark, stems, and leaves are believed to be antidiabetic and are administered to lower blood sugar levels. In some of the Caribbean islands, a leaf decoction is taken as a remedy for diarrhea, fever, chest complaints, diabetes, and hypertension. In Suriname a mango leaf infusion is used to lower hypertension and promote good blood circulation. In Nigeria the leaves are used as an antidiabetic agent. The tender leaves are cooked and eaten as a vegetable in Indonesia and the Philippines. Various plant parts of the mango are used in a wide spectrum of household remedies in India and other countries. The ash of burned leaves is a remedy for burns and scalds, and fumes from burning leaves give relief from hiccups and infection of the throat (Majumdar and Sharma 1985). Fresh leaves are masticated to tone up the gums. The bark acts as a tonic on mucous membrane. Flowers have curative properties, and cooked fruits are used to treat diarrhea and chronic dysentery (Sood and Kapil 1990). The kernel juice is snuffed once a day for three days to stop nasal bleeding. The unripe fruits are useful in ophthalmia and skin eruptions. The rind is a stimulating tonic in debility of the stomach (Chunekar 1999).

The mango is a large, erect, evergreen tree growing to 10 to 30 m (30-100 ft) with a dark green, spreading, umbrella-shaped crown. The leaves are alternate, simple, leathery, lanceolate, and 18 to 30 cm (6-12 in) long. The leaves have long petioles and are arranged in whorls at the tips of shoots. The midrib is pale and conspicuous. The inflorescence is a panicle approximately 45 cm long, yellowish green, and a bearer of numerous flowers. The five-petal flowers are off-white and streaked. The fruits grow at the end of long stringlike flower stalks (peduncles). The fruit size, shape, and color varies widely depending on the cultivar. The outer skin is thick and glandular. The mango fruit is a large, fleshy drupe, containing edible flesh of varying thickness. The flesh of the fruit can be fibrous or nonfibrous and taste sour to sweet depending on the cultivar. The seed kernel is inside a thick woody covering. The young and the unripe fruits of mango are acidic in taste and used for various culinary purposes. The ripe fruits are used in preparing various processed products such as fruit drinks, nectar, jam, cereal flakes, custard powder, baby food, and toffee (CSIR 1962).

Mangoes are an important part of the diet in many developing countries in the tropics and subtropics as a source of vitamins A and C, calcium, phosphorus, and iron. The fruit is eaten as dessert or sliced in fruit salad, dry cereal, gelatin, custards, or ice cream. The ripe flesh may be spiced and preserved in jars, sliced and canned in syrup, or made into jam, marmalade, jelly, or nectar. The extracted pulpy fruit juice is used for making mango halva, desserts, and drinks. Mango juice is spray-dried and powdered and used in invalid and baby foods.

Phytochemical research of different parts of *M. indica* has demonstrated the presence of phenolics, triterpenes, flavonoids, phytosterols, and polyphenols (Anjaneyulu et al. 1994; Kharn et al. 1994; Saleh and El-Ansari 1975; Selles et al. 2002; Singh et al. 2004). The mango is a rich source of vitamins A and C. The ascorbic acid content varies from 3.2 to 62.3 mg/100 g pulp (CSIR 1962). The presence (in mg/100 g FW) of folic acid (vitamin B; 3.6), vitamin B_1 (thiamine; 35-63), and vitamin B_2 (riboflavin; 37-73) in raw mango has been reported by Majumdar and Sharma (1985). Ripe fruits contain carotenoid pigments, beta-carotene (provitamin A), 7.09% to 17.2% total soluble sugar, 0.2% to 0.5% titratable acids, 1.0% protein, and 11.8% to 26.8% total soluble solids. Sugars in mangoes include sucrose, glucose, fructose, and maltose. The presence of amino acids, including the essential lysine, leucine, cysteine, valine, arginine, phenylalanine, and methionine, in fruits of various cultivars has been recorded (Majumdar and Sharma 1985). Concentration of amino acids varies from 10 to 40 mg/ 100 g edible pulp.

Mangoes have been reported to contain high amounts of carotenoids (Chen et al. 2004; Setiawan et al. 2001) and soluble dietary fiber (Ramulu and Rao 2003). Carotenoids are well known for improvement of human health (Rao and Agarwal 1999; Sesso et al. 2004). For instance, beta-carotene has been shown to possess high vitamin A activity and antioxidant capacity (Mercadante and Rodriguez-Amaya 1998; Miller et al. 1996). Codoy and Rodriguez-Amaya (1989) reported that beta-carotene is the dominant carotenoid in mangoes, comprising 48% to 84% of the total. The composition of carotenoids in mangoes can be affected by many factors, such as the cultivar, geographic origin, the fruit's degree of maturity, and processing conditions (Cano and de Ancos 1994; Pott et al. 2003).

The composition of mango kernels and their utilization as a source of fat (Gaydou 1984; Lakshminarayana et al. 1983; Moharram and Moustafa 1982), natural antioxidants (Puravankara et al. 2000), starch (Kaur et al. 2004), flour (Arogba 2002), and feed (Omoregie 2001; Ravindran and Sivakanesan 1996) have been investigated in a number of studies. Mango peel as a source of pectin has been proposed by several authors (Beerh et al. 1976; Pedroza-Islas et al. 1994; Srirangarajan and Shrikhande 1976; Tandon and Garg 1999; Tandon et al. 1991); however, only a small number of cultivars have been studied with respect to contents and properties of the pectin (Beerh et al. 1976; Kratchanova et al. 1991; Pedroza-Islas et al. 1994; Srirangarajan and Shrikhande 1976, 1977, 1979; Sudhakar and Maini 2000; Tandon and Garg 1999; Tandon et al. 1991). Mango peels contain a number of flavonol *O*- and xanthone *C*-glycosides (Schieber et al. 2003). These compounds might be used as natural antioxidants to replace synthetic additives or as functional food ingredients.

Several groups of compounds contribute to the important medicinal properties of mangoes. For instance mangiferin, a tetrahydroxy pyrrolidine saponin, has been isolated from the leaves of mangoes and in some cases showed antiviral effects (Zhu et al. 1993). Mangiferin (a *C*-glucosylxanthone) has cardiotonic, diuretic, and antioxidant properties and prevents lipoperoxidation (Ghosal 1996). Mangiferin has been demonstrated to exhibit anti-inflammatory, hypoglycemic, and antidiabetic activity by lowering the intestinal absorption of glucose (Aderibigbe et al. 1999, 2001). Aqueous extract from mango tree bark exhibited antinociceptive, anti-inflammatory, and antioxidant actions (Garrido et al. 2001; Martinez et al. 2000; Sanchez et al. 2000; Scartezzini and Speroni 2000). Other components of mangoes, gallic acid and quercetin, are capable of strong antiviral activities (Rastogi and Mehrotra 1995). Stem-bark extract of mango showed antiamoebic activity and was effective against diarrhea (Tona et al. 1998). Mango kernel powder was shown to have antimicrobial effects against a number of food-borne pathogenic bacteria (Kabuki et al. 2000). Consumption of mango reduced the risk of gallstones and gallbladder cancer (Pandey and Shukla 2002).

In 2005 global mango production was 27,966,749 Mt and the area harvested was 3,870,200 ha (FAOSTAT 2006).

Banana (Musa *spp.; Musaceae)*

Bananas are native to tropical Asia. The term *banana* embraces a number of species or hybrids in the genus *Musa* of the family Musaceae. Almost all known edible-fruit cultivars arose from two diploid species, *M. acuminata* and *M. balbisiana,* which are native to Southeast Asia. More than 400 varieties of bananas are known. The edible banana is the fruit of a sterile hybrid plant. The plantain, a larger hybrid variety used more as a green vegetable, is also a dietary staple in many countries. Banana is the fourth-largest fruit crop of the world. Bananas and plantains are grown in several tropical regions of the world, including Latin America, Southeast Asia, and Africa.

The banana plant grows to 4 to 9 m (13-30 ft) with leaf sheaths forming a trunklike pseudostem. The large leaves grow up to 2.7 m (9 ft) long and 0.6 m (2 ft) wide. The true underground stem grows from the heart of the pseudostem and bears the inflorescence at the terminal end. Flowers develop in clusters and around the main axis of the stem. The female flowers produce edible fruits without fertilization.

One hundred grams of banana fruit pulp can supply up to 24 g carbohydrates, 11 mg calcium, 28 mg phosphorus, and 241 mg potassium. Bananas contain 12.7 mg/100 g vitamin C (ascorbic acid), 96.9 µg/100 g beta-carotene (provitamin A), and 104.9 µg/100 g alpha-carotene. The average potassium content for Hawaii's bananas is 330.6 mg/100 g (Wall 2006a).

The fruits are eaten raw; processed as pulp; pureed for use in a variety of products including ice cream, yogurt, cake, bread, honey, and baby food; sliced and canned in syrup; candied and used in baking, fruit salads, and toppings; deep-fried as chips; or dried and preserved with jaggery (a type of molasses). Banana fruit extract flavors beverages and dairy products. Green fruits can also be dried and ground into flour. Unripe bananas of selected varieties are sliced and deep-fried to make chips that are favorite snack items of Asian countries. Fermented ripe bananas are also used to make vinegar and alcoholic beverages.

The heart of the pseudostem and the male bud of the banana plant are consumed as vegetables in India and Southeast Asia. All parts of the banana plant have medicinal properties and are used in Ayurvedic medicine—fresh plantain sap is administered orally as a purgative; flower extracts treat bronchitis, dysentery, ulcers, diabetes, and gynecologic problems; and aqueous plant extracts treat dysentery, diarrhea, and insect bites. Young leaves of the banana plant are used as disinfectant poultices to treat burns, and stem extracts treat kidney stones.

Several oligosaccharides, including fructose, xylose, galactose, glucose, and mannose, occur naturally in bananas, making them an excellent prebiotic for the selective growth of beneficial bacteria in the intestine (Gibson 1998). Banana flower extract has a hypoglycemic effect (Dhanabal et al. 2005; Pari and Maheswari 1999, 2000). Banana peel and pulp contain high concentrations of dopamine, which has high antioxidant properties (Kanazawa and Sakakibara 2000). Banana fruit pulp to treat ulcers has been explored by a number of investigators (Best et al. 1984; Elliot and Heward 1976; Goel et al. 1989; Lewis and Shaw 2001).

In 2005 global banana production was 72,624,562 Mt and the area harvested was 4,456,485 ha (FAOSTAT 2006).

*Litchi (***Litchi chinensis** *Sonn.; Sapindaceae)*

The litchi is a tropical fruit that originated in southern China. It has a bright red, attractive pericarp surrounding a white aril (Nakasone and Paull 1998). The litchi tree grows to about 12 m (40 ft) and has leathery foliage. The leaves are pinnate and divided into four to eight leaflets. They are reddish when young, becoming shiny and bright green. The flowers are small, yellowish green, and borne in terminal clusters. The fruit is covered by a leathery rind, pink to red, rough, and oval. The edible, aril portion is white, translucent, firm, and juicy. It has a fragrant and sweet flavor.

Pharmacological studies show that leaf extracts are nontoxic, anti-inflammatory, analgesic, and antipyretic (Besra et al. 1996). The fruit pericarp contains significant amounts of polyphenolic compounds and flavonoids, which exhibit powerful antioxidant activity against fat (Zheng et al. 2003) and potential anticancer activity (Wang et al.

2005, 2006). Gorinstein et al. (1999) reported that litchis have 3.35 mg total polyphenols in 100 g fresh fruit. Wall (2006b) found (in mg/100 g FW) ascorbic acid, 36; phosphorus, 31; potassium, 181; calcium, 5; magnesium, 16; sodium, 8; iron, 0.43; zinc, 0.28; manganese, 0.11; and copper, 0.23. Wall also documented that litchi fruit consumption would supply 2% to 4% of the dietary reference intake (DRI) for six minerals (P, K, Mg, Fe, Zn, Mn) and provide 22% of the DRI for copper. Consuming 14 to 17 litchis meets the DRI for vitamin C for the average adult (75-90 mg; Wall 2006b). Numerous studies have documented the presence of large amounts of anthocyanins in the litchi pericarp (up to 0.20 mg/g FW), giving the litchi good antioxidant potential (Lee and Wicker 1991; Sarni-Manchado et al. 2000; Zhang et al. 2000a,b; Zhang et al. 2004).

*Rambutan (*Nephelium lappaceum; *Sapindaceae)*

Although relatively unknown in the United States, the Malayan rambutan is an important commercial crop in Asia, where it is consumed fresh, canned, or processed and appreciated for its refreshing flavor and exotic appearance (Almeyada et al. 1979).

The tree is a tall evergreen reaching 12 to14 m (40-45 ft) and has a dense, spreading crown. The leaves are alternate, pinnately compound, and hairy when young, and have one to four pairs of leaflets, slightly leathery, on a reddish rachis. The small white flowers are borne in axillary, branched, hairy panicles. The fruit is ovoid or ellipsoid, pinkish red to maroon or dark purple, and thin and has a leathery rind covered with tubercles from each of which extends a soft fleshy red, pinkish, or yellow spine 0.5 to 2 cm long, the tips deciduous in some types. Inside the rind the white or slightly pinkish translucent, juicy, sweet pulp is 0.5 to 1.0 cm thick, adhering more or less to the ovoid or oblong, somewhat flattened seed, which is 2.5 to 3.5 cm long and 1 to 1.5 cm wide.

Rambutan fruits are a good source of copper and manganese and provide 2% to 6% of the DRI for five minerals: phosphorus, potassium, magnesium, iron, and zinc. The DRI for vitamin C for the average adult (75-90 mg) can be met by consuming 10 to 12 rambutans (Wall 2006b). Wall (2006b) found that rambutans contain (in mg/100 g FW) ascorbic acid, 48; phosphorus, 19; potassium, 250; calcium,

8.8; magnesium, 17.2; sodium, 8.2; iron, 0.56; zinc, 0.26; manganese, 0.38; and copper, 0.2.

*Longan (*Dimocarpus longan; *Sapindaceae)*

The longan is native to southern China. The plant is a large evergreen tree growing up to 12 m (40 ft.) tall and spreading about 14 m (45 ft) wide. The leaves are alternate, paripinnate elliptic, ovate-oblong or lanceolate with a blunt tip, leathery, slightly pubescent, and glossy green on the upper surface and grayish beneath. The flowers are pale yellow, and the globose fruits are drooping clusters with thin, brittle, yellow brown to light reddish brown rinds.

Longans are a good source of potassium (324.9 mg/100 g FW) and copper (0.26 mg/100 g FW), and they provide 3% to 5% the DRI of phosphorus, magnesium, iron, manganese, and zinc. Wall (2006b) found that longans contain (in mg/100 g FW) ascorbic acid, 80; phosphorus, 36; potassium, 338; calcium, 10.7; magnesium, 14.3; sodium, 18.7; iron, 0.67; zinc, 0.3; manganese, 0.1; and copper, 0.32. The DRI for vitamin C for the average adult (75-90 mg) can be met by consuming about 12 to 14 longans (Wall 2006b).

Longan fruits are extensively consumed in Thailand and other tropical countries in Asia, Africa, and Central and South America. In Chinese medicine the flesh of the longan is used as a stomachic, antipyretic, anthelmintic and antidote for poisons (Mortin 1987). A decoction of the dried flesh is used as a tonic and to treat insomnia and neurasthenic neurosis (Mortin 1987). The seed and other fruit components contain adenosine, which may account for the longan's reported sedative and analgesic effects (Okuyama et al. 1999), and corilagin (Cheng et al. 1995; Hsu et al. 1994), which is studied for its pharmacological activities by itself and in combination with other plants (Notka et al. 2003). Corilagin is hypotensive (Cheng et al. 1995), antifungal against *Candida glabrata* strains (Latte and Kolodziej 2000), potently inhibits HIV (Notka et al. 2003), and inhibits tumor necrosis (Okabe et al. 2001); it is also an antibiotic (Shiota et al. 2004).

*Tamarind (*Tamarindus indica; *Caesalpiniaceae)*

The tamarind originated in tropical east Africa, India, and some parts of Southeast Asia. It is mentioned in Indian Brahmasamhita scriptures written between 1200 and 200 BC, in Buddhist writings from AD 650, and in Marco Polo's writings from 1298. India is the primary producer and exporter of tamarind, which is also cultivated in China, Pakistan, Africa, Indochina, the Philippines, Java, Spain, Southeast Asia, Australia, and the southern United States.

The pulp of the fruit is used for seasoning, as a food component, and in juices. Tamarind pulp is used to make soft drinks in Latin America, Jamaica, and the Middle East. Tamarind pulp is used extensively in Indian and Southeast Asian cuisine, the fruits are used to make drinks and beverages, and it is an essential ingredient in sauces to impart sourness. The tamarind's role as an antibacterial, antifungal, and food preservative is recognized by households in India. Tamarind pods are available commercially as preserved pods, pulp, and dried powder. The dried pods are used to make chutneys, curries, and sauces. The juice is used to pickle fish and vegetables in India. The soft, succulent pulp is used either fresh or dried as a confectionery and as an ingredient in curries, chutneys, preserves, pickles, sherbets, and beverages. It is also candied in some Asian countries. The tender pods are sometimes cooked or pickled and eaten as a vegetable. The seeds are extracted from pods, and the pulp is stored for several months in a compressed form in huge ceramic jars in India. In Japan purified and refined polysaccharide from hulled tamarind seed is used in commercial food products as a gelling, thickening, and stabilizing agent (Gidley et al. 1991).

In India and Southeast Asia, tamarind seeds are crushed and boiled to produce a paste that is used as a roofing material. Several medicinal uses of tamarind have been reported. Tamarind is used in traditional medicine as a cathartic, astringent, antipyretic, antiseptic, and a preservative. The fruit is said to reduce fever, cure intestinal ailments, and be effective against scurvy. Infusions and decoctions from the leaves are used as an anthelmintic and for treating jaundice. In some parts of India and Africa it is used to treat dysentery. The fruit is said to improve digestion, relieve gas, soothe sore throats, and act as a mild laxative. In Suriname, the bark is used to treat diarrhea and the boiled leaves are used to treat skin disorders, such as scabies.

Ripe tamarind fruit has numerous well-recognized and proven medicinal values (Rao et al. 1999). The fruit extracts exhibit anti-inflammatory activities (Rimbau et al. 1999). Its fruit is regarded as a digestive, carminative, laxative, expectorant, and blood tonic (Komutarin et al. 2004). Other parts of the plant have antioxidant, hepatoprotective

(Joyeux et al. 1995), anti-inflammatory, antimutagenic (Ramos et al. 2003), and antidiabetic (Maiti et al. 2004) activities. Studies have substantiated the traditional use of tamarind as a food preservative, disinfectant, and antiseptic (De et al. 1999). Yi et al. (2005) reported the nutritional value and chemical constituents of three cultivars: total sugar content ranged from 45.90% to 50.26%; reducing sugar, 33.66% to 46.94%; and ascorbic acid, 0.33% to 0.87%; calcium content was 1.2 to 2.0 mg/g, and there was 27.68% essential amino acids relative to 45.85% total amino acids.

The antimicrobial action of tamarind leaves and pulp (Alian et al. 1983; Solis 1969) and seed extracts against several fungi and bacteria, including *Escherichia coli,* is due to the chemical compound tamarindineal (5-hydroxy-2-oxo-hexa-3, 5-dienal; Imbabi et al. 1992). Tamarind has been studied for treating obesity (Sergio 1998). Crude extracts from the tamarind seed coat exhibit antioxidant activity, suggesting that the tamarind seed coat, a by-product of tamarind gum industries, can be used as a safe and low-cost source of antioxidants (Tsuda et al. 1994, 1995). A polysaccharide isolated and purified from tamarinds showed immunomodulatory activities such as phagocytic enhancement, leukocyte migration inhibition, and lymphocyte proliferation inhibition, suggesting that polysaccharide from the tamarind may aid in human health promotion and disease prevention (Sreelekha et al. 1993). Other studies have shown the potential of tamarind extracts to diminish the risk of atherosclerosis development in humans (Martinello et al. 2006).

*Jackfruit (***Artocarpus heterophyllus;** *Moraceae)*

Jackfruit is indigenous to the tropical Western Ghats of India. It is the largest tree-borne fruit in the world. Jackfruit is famous as one of the royal trinity of fruits in Tamil culture (mango, banana, and jackfruit) in India and mentioned in ancient literature and folklore. The delicious fruit is popular in the summer months, when it is harvested and available as a local favorite of the season. It substitutes in the diet of Bangladesh as a vegetable and as a nutritious food during the summer season, appropriately named the poor man's food in Southeast Asia. Jackfruit is an economically important crop as well as a backyard tree in several tropical countries, including India, Southeast

Asia, the East Indies, eastern Africa, Brazil, and the Philippines. Jackfruit is one of the major tropical fruits of the world.

The jackfruit tree is large, growing to 12 m (40 ft) tall, with a strong trunk and large, deeply lobed, alternate, glossy, leathery leaves. The tree is monoecious, with tiny male flowers borne in oblong clusters and elliptic or rounded female flowers. The fruit is an aggregate fruit with a hard rind. The rind is composed of numerous hard, conelike points attached to a thick, rubbery, pale yellow or whitish wall. The interior consists of fully developed yellow perianths (which form the edible portion of the fruit), embedded in masses of undeveloped perianths and a central, pithy core. Each developed perianth encloses an oval, light brown seed enclosed in a rind of its own.

The fruits are harvested when completely mature, which is normally judged by the size and a slight color change of the skin. After the fruits are harvested, the fruit-bearing twigs are cut back to the trunk or the branch to induce flowering in the next season. Although approximately 60% of the fruit is nonedible, almost all parts of it are used. Tender young fruits are sometimes made into pickles or cooked as a vegetable.

The ripe fruit is an excellent table fruit by itself and is used to make numerous desserts, chips, and other snack foods. The fruits are often enjoyed with ghee or coconut milk in India; used to make ice cream, chutney, jam, jelly, or paste; or canned in syrup made with sugar or honey. Fresh fruits are used as a daily fruit in the Western Ghats and as a major ingredient in numerous desserts and baby food. The fruits are also dried or frozen for storage. The nonedible parts of the jackfruit constitute the waste, which is a rich source of carbohydrates, fat, protein, minerals, and amino acids and is useful as cattle feed. The seeds are starchy and used as a vegetable, roasted, baked, or boiled in curried dishes. In India the leaves are used as wrappers in cooking and are often sewn together for use as plates.

Jackfruit is also listed as a medicinal plant in Chinese and Indian medicine. The Chinese believe that jackfruit pulp and seeds have tonic, cooling, and nutritious properties, and they consider the roasted seeds an aphrodisiac. The ashes of charred jackfruit leaves are used to heal ulcers; the latex is used to heal abscesses, snakebite, and glandular swellings; and the tree root is used to treat skin diseases, asthma, fever, and diarrhea. Ayurveda considers jackfruit as being cool and destroying excessive heat, bile, and wind in the human body. It is also proposed that jackfruit develops flesh, phlegm, and strength in the body and increases virility. The juice extracted from both jackfruit and mango tree bark when mixed with limestone water is prescribed for diarrhea.

Jackfruit is a good source of ascorbic acid and carotene (Hossain et al. 1979). A lectin in jackfruit seeds, jacalin (Chatterjee et al. 1979), has gained importance. Jacalin has the unique property of binding human IgA and has emerged as an important tool with diverse applications in biomedical studies, such as isolation of human plasma glycoproteins, investigation of IgA nephropathy and detection of tumors (Kabir 1998; Sankaranarayanan et al. 1996; Sujathan et al. 1996), general diagnosis of oral cancers (Remani et al. 1997), and differential diagnosis of premalignant and malignant lesions of the oral cavity (Vijayan et al. 1987).

Mangosteen (**Garcinia mangostana**; *Clusiaceae*)

Mangosteen is native to Southeast Asia, probably the Malay Archipelago, particularly the Sunda Islands and Moluccas. It is cultivated in tropical rain forests in Indonesia, Malaysia, the Philippines, and Thailand. It is named the queen of fruits because it is considered the best-tasting fruit in the world. It is a slow-growing tropical evergreen tree with leathery, glabrous leaves. The fruit's hard skin is dark purplish brown. It is the size and shape of a tangerine, having five to eight fruit segments. The edible fruit aril is white, soft, and juicy with a sweet, slightly acidic taste and a pleasant aroma (Martin 1980). The segmented flesh is soft, cream colored, and juicy. It has an extremely refreshing sweet-tart flavor (Herbst 2001). It is marketed as a dietary supplement in juice or capsule form. The juice consists of the whole fruit of the mangosteen pureed with other juices, such as blueberry, cranberry, raspberry, grape, cherry, strawberry, pear, and apple.

The pericarp of mangosteen has been used in Thai indigenous medicine to treat skin infections, wounds, and diarrhea for many years (Martin 1980; Mahabusarakam et al. 1987; Moongkarndi et al. 2004). It is also used to treat abdominal pain, dysentery, suppuration, chronic ulcer, leukorrhea, and gonorrhea and as an astringent (Satyavati et al. 1976). The nutritional content (per 100 g) is 24 g total carbohydrate, 16 g sugar, and 560 mg potassium (XanGo 2006). Products manufactured from mangosteen are used in the United States as a dietary supplement because of their antioxidant potential (Garrity et al. 2005). The major secondary metabolites of mangosteen have been found to be xanthone derivatives (Jung et al. 2006; Mahabusarakam et al. 1987; Nilar et al. 2005; Suksamrarn et al. 2002, 2003), which possess antifungal (Gopalakrishnan et al. 1997), antimicrobial (Suksamrarn et al. 2003), antioxidant (Yoshikawa et al. 1994), and cytotoxic (Ho et al. 2002) activities. Mangosteen extracts have a potential for neuroprotection in oxidative stress-induced cell damage in neurodegenerative diseases such as Alzheimer disease, Parkinson disease, and stroke (Weecharangsan et al. 2006). Others demonstrated anti-inflammatory (Gopalakrishnan et al. 1980), antitumor, antioxidant (Williams et al. 1995), and antibacterial activities (Farnsworth and Bunyapraphatsara 1992; Mahabusarakum et al. 1983). The peel contains mangostin, tannin, xanthone, chrysanthemin, garcinone, gartanin, vitamins B_1, B_2, C, and other bioactive substances (Farnsworth and Bunyapraphatsara 1992). The peel is antiproliferative against cancer cell lines, shows potent antioxidant activities, and induces apoptosis in breast cancer cell lines (Moongkarndi et al. 2004).

*Citrus Fruits (*Citrus *spp.; Rutaceae)*

The genus *Citrus* consists of many orange and lemon types. Sweet oranges (*C. sinensis*) have at least four subclasses: common, navel, Valencia, and blood. Tangerine species are *C. reticulata, C. clementina, C. deliciosa, C. nobilis,* and *C. unshiu.* Tangors are orange-tangerine hybrids (*C. reticulata* × *C. sinensis*). Tangelos are tangerine-grapefruit or pummelo hybrids (*C. reticulata* × *C. paradisi* or *C. reticulata* × *C. grandis*). The sour orange is *C. aurantium;* lemon, *C. limon;* and lime, *C. aurantifolia.*

Citrus fruit consumption appears to be associated with lower risk of colorectal (Levi et al. 1999), esophageal (Chen et al. 2002; Levi et al. 2000), gastric (Palli et al. 2001), and stomach (McCullough et al. 2001) cancers as well as stroke (Feldman 2001; Joshipura et al. 1999). Citrus fruits also appear to be associated with improved blood lipid profiles (Kurowska et al. 2000) and improved survival in the elderly (Fortes et al. 2000). Citrus phytochemistry has been widely studied, particularly flavonoids for their functional properties (Areias et al. 2001; Bae et al. 1999; Bear and Teel 2000a,b; Borradaile et al.

1999; de Gregorio et al. 2000; Jeon et al. 2001; Kato et al. 2000; Kim et al. 2000; Kohno et al. 2001; Lee et al. 2001; Manthey et al. 2001; Miyake et al. 2000; Wilcox et al. 2001; Zhang et al. 2000b). These studies indicate that citrus fruits could be notable sources of antioxidant, anti-inflammatory, and cancer-preventive compounds (Murakami et al. 2000).

The juice yield of citrus fruits is about half the fruit weight (Bovill 1996), and the rest of the fruits generate very large amounts of by-products. Peel and seed residue is the primary waste fraction. Peels are a source of molasses, pectin, cold-pressed oils, and limonene and can be used as cattle feed when mixed with dried pulp. Seeds are rich in unsaturated fatty acids and can be used to recover limonoids, which are typical citrus fruit triterpenoids having an extremely bitter taste and probably anticarcinogenic or chemopreventive activities (Braddock 1995). Limonoids are secondary metabolites present in all citrus fruit tissues; studies have shown they inhibit human cancer cell proliferation and induce apoptosis of human breast cancer cells (Tian et al. 2001; Satoru et al. 1999; Vanamala et al. 2004).

Both peels and seeds are rich sources of phenolic compounds, which include phenolic acids and flavonoids. Flavonoids are represented in citrus fruits by polymethoxylated flavones and glycosylated flavanones. Flavonoids are found only in citrus fruits, and each species has a specific pattern, which makes them very good markers to determine adulteration in commercial juices (Marini and Balestrieri 1995; Mouly et al. 1994; Ooghe and Detavernier 1997). The citrus flavonoids have been found to have anticancer, antiviral, and anti-inflammatory activities; the ability to decrease capillary fragility; and an effect that inhibits platelet aggregation (Benavente-Garcia et al. 1997; Huet 1982). Some glycosylated flavanones are potent natural sweeteners (Bör et al. 1990; Horowitz and Gentili 1969). Several studies have shown the antioxidant activity in food systems of several citrus fruits (sweet orange, lemon, grapefruit), used directly (Piskur and Higgins 1949; Williams and Harris 1983) or as extracts (Kroyer 1986; Pereira and Mancini-Filho 1994; Sawamura et al. 1988; Ting and Newhall 1965).

Oranges (*C. sinensis*) are antiproliferative against cancer cell lines (Satoru et al. 1999). Blood oranges are characterized by their unique flesh and rind color due to phenolic pigments belonging to the anthocyanins. Blood orange fresh tissue and juice contain a predominance of anthocyanins (e.g., cyanidin-3-glucoside and cyanidin-3-(6″-malonyl)-glucoside; Dugo et al. 2003; Maccarone et al. 1998). Rapisarda et al. (1999) reported in vitro studies that showed the antioxidant property of blood orange juice was due more to anthocyanin rather than vitamin C content.

REFERENCES

Aderibigbe AO, Emudianughe TS, Lawal BA. 1999. Antihyperglycaemic effect of *Mangifera indica* in rat. Phytotherapy Research 13(6): 504-507.

Aderibigbe AO, Emudianughe TS, Lawal BA. 2001. Evaluation of the antidiabetic action of *Mangifera indica* in mice. Phytotherapy Research 15(5): 456-458.

Alian A, El-Ashwah E, Eid N. 1983. Antimicrobial properties of some Egyptian nonalcoholic beverages with special reference to tamarind. Egyptian Journal of Food Science 11(1-2): 109-114.

Almeyada N, Mab SE, Martin FW. 1979. The rambutan. Citrus Sub-Tropical Fruit Journal 544: 10-12.

Anjaneyulu V, Babu IS, Connollu JD. 1994. 29-Hydroxymangiferonic acid from *Mangifera indica*. Phytochemistry 35: 1301-1303.

Areias FM, Rego AC, Oliveira CR, Seabra RM. 2001. Antioxidant effect of flavonoids after ascorbate/Fe(2+)-induced oxidative stress in cultured retinal cells. Biochemical Pharmacology 62: 111-118.

Arogba SS. 2002. Quality characteristics of a model biscuit containing processed mango (*Mangifera indica*) kernel flour. International Journal of Food Properties 5: 249-260.

Bae EA, Han MJ, Kin DH. 1999. *In vitro* anti-*Helicobacter pylori* activity of some flavonoids and their metabolites. Planta Medica 65: 442-443.

Bak I, Lekli I, Juhasz B, Nagy N, Varga E, et al. 2006. Cardioprotective mechanisms of *Prunus cerasus* (sour cherry) seed extract against ischemia/reperfusion-induced damage in isolated rat hearts. American Journal of Physiology—Heart and Circulatory Physiology 291: H1329-H1336.

Bear WL, Teel RW. 2000a. Effects of citrus phytochemicals on liver and drug cytochrome P450 activity and on the *in vitro* metabolism of the tobacco-specific nitrosamine NNK. Anticancer Research 20: 3323-3329.

Bear WL, Teel RW. 2000b. Effects of citrus flavonoids on the mutagenicity of heterocyclic amines and on cytochrome P450 1A2 activity. Anticancer Research 20: 3609-3614.

Beerh OP, Raghuramaiah B, Krishnamurthy GV. 1976. Utilization of mango waste: Peel as a source of pectin. Journal of Food Science and Technology 13: 96-97.

Benavente-Garcia O, Castillo J, Marin FR, Ortuno A, Del Rio JA. 1997. Uses and properties of Citrus flavonoids. Journal of Agricultural and Food Chemistry 45: 4505-4515.

Besra SE, Sharma RM, Gomes A. 1996. Antiinflammatory effect of petroleum ether extract of leaves of *Litchi chinensis Gaertn* (Sapindaceae). Journal of Ethnopharmacology 54: 1-6.

Best R, Lewis DA, Nasser N. 1984. The antiulcerogenic activity of the unripe plantain banana (*Musa* species). British Journal of Pharmacology 82: 107-116.

Blando F, Gerardi C, Nicoletti I. 2004. Sour cherry (*Prunus cerasus* L) anthocyanins as ingredients for functional foods. Journal of Biomedicine and Biotechnology 5: 253-258.

Bomser J, Madhavi DL, Singletary K, Smith MA. 1996. *In vitro* anticancer activity of fruit extracts from *Vaccinium* species. Planta Medica 62(3): 212-216.

Bör A, Borrego F, Benavente O, Castillo J, Del Rio JA. 1990. Neohesperidin dihydrochalcone: Properties and applications. Lebensmittel-Wissenschaft und - Technologie 23: 371-376.

Borradaile NM, Carroll KK, Kurowska EM. 1999. Regulation of HepG2 cell apolipoprotein B metabolism by the citrus flavanones hesperetin and naringenin. Lipids 34: 591-598.

Bovill H. 1996. L'orange: Source de molécules naturelles [Orange: Source of natural compounds]. Arômes Ingrédients Additifs 7: 41-42.

Braddock RJ. 1995. Byproducts of citrus fruits. Food Technology 49: 74-77.

Cano MP, de Ancos B. 1994. Carotenoid and carotenoid ester composition in mango fruit as influenced by processing method. Journal of Agricultural and Food Chemistry 42: 2737-2742.

Chatterjee BP, Vaith P, Chaterjee S, Karduck D, Uhlenbruck G. 1979. Comparative studies of new marker lectins for alkali-labile and alkali-stable carbohydrate chains in gllycoproteins. International Journal of Biochemistry 10: 321.

Chen H, Ward MH, Graubard BI, Heineman EF, Markin RM, et al. 2002. Dietary patterns and adenocarcinoma of the esophagus and distal stomach. American Journal of Clinical Nutrition 75: 137-144.

Chen JP, Tai CY, Chen BH. 2004. Improved liquid chromatographic method for determination of carotenoids in Taiwanese mango (*Mangifera indica* L.). Journal of Chromatography A 1054: 261-268.

Cheng JT, Lin TC, Hsu FL. 1995. Antihypertensive effect of corilagin in the rat. Canadian Journal of Physiology and Pharmacology 73: 1425-1429.

Chopra RN, Nayar SL, Chopra IC. 1956. Glossary of Indian medicinal plants. CSIR, New Delhi, India. p. 2040.

Chunekar KC. 1999. *Mangifera indica* L. (Mango). In: Pandey GC, editor. Bhavaprakas nighantu (Indian materia medica of Sri Bhavamishra c. 1500-1500 AD). Bharati Academy, Chaukhambha Varanasi. p. 550-554.

Codoy HT, Rodriguez-Amaya DB. 1989. Carotenoid composition of commercial mangoes from Brazil. Lebensmittel-Wissenschaft and Technologie 22: 100-103.

CSIR (Council of Scientific and Industrial Research). 1962. Raw materials. In: The Wealth of India, A Dictionary of Indian Raw Materials and Industrial Products. Vol. 6(L-M). Publications and Information Directorate, CSIR, New Delhi, India. p. 262-285.

Cui T, Nakamura K, Ma L, Li JZ, Kayahara H. 2005. Analyses of arbutin and chlorogenic acid, the major phenolic constituents in Oriental pear. Journal of Agricultural and Food Chemistry 53(10): 3882-3887.

De M, Krishna DA, Banerjee AB. 1999. Antimicrobial screening of some Indian spices. Phytotherapy Research 13(7): 616-618.

De Gregorio AC, Garcia-Domenech R, Galvez J, Ros MJ, Wolski S, et al. 2000. Molecular topology: A useful tool for the search of new antibacterials. Bioorganic and Medicinal Chemistry Letters 10: 2033-2036.

Dhanabal SP, Sureshkumar M, Ramanathan M, Suresh B. 2005. Hypoglycemic effect of ethanolic extract of *Musa sapientum* on alloxan induced diabetes mellitus in rats and its relation with antioxidant potential. Journal of Herbal Pharmacotherapy 5(2): 7-19.

Ding T. 1983. Experimental study of microcirculatory changes in asthma and its prevention in guinea pigs by traditional Chinese herbs. Chinese Journal of Integration of Traditional and Western Medicine 3: 37-38.

Dugo P, Mondello L, Morabito D, Dugo G. 2003. Characterization of the anthocyanin fraction of Sicilian blood orange juice by micro-HPLC-ESI/MS. Journal of Agricultural and Food Chemistry 51: 1173-1176.

Duke JA. 1992. *Prunus persica;* Peach. In: Handbook of phytochemical constituents of GRAS herbs and other economic plants. CRC Press, Tokyo. p. 488-490.

Einbond LS, Reynertson KA, Luo XD, Basile MJ, Kennelly EJ. 2004. Anthocyanin antioxidants from edible fruits. Food Chemistry 84: 23-28.

Elliott RC, Heward GJ. 1976. The effect of banana-supplemented diet on gastric ulcer in mice. Pharmacological Research Communications 8: 167-171.

FAOSTAT (Statistics Division of the Food and Agriculture Organization of the United Nations). 2006. FAO Statistical Database-Agriculture. http://faostat.fao .org. Accessed February 2006.

Farnsworth NR, Bunyapraphatsara N. 1992. Thai medicinal plant: Recommended for primary health care system. Prachachon, Bangkok.

Feldman EB. 2001. Fruits and vegetables and the risk of stroke. Nutrition Reviews 59: 24-27.

Fortes C, Forastiere F, Farchi S, Rapiti E, Pastori G, et al. 2000. Diet and overall survival in a cohort of very elderly people. Epidemiology 11: 440-445.

Garrido G, Gonzalez D, Delporte C, Backhouse N, Quintero G, et al. 2001. Analgesic and anti-inflammatory effects of *Mangifera indica* L. extract (Vimang). Phytotherapy Research 15(1): 18-21.

Garrity AR, Morton JC, Morrison P, De La Huerga V. 2005. Nutraceutical mangosteen tea. PCT Int Appl WO 2005048940, A2 20050602. p. 9.

Gaydou EM. 1984. Sterols, methyl sterols, triterpene alcohols and fatty acids of the kernel fat of different Malagasy mango (*Mangifera indica*) varieties. Journal of the American Oil Chemists' Society 61: 1589-1593.

Ghosal S. 1996. A plausible chemical mechanism of bioactivities of mangiferin. Indian Journal of Chemistry 35B: 561-566.

Gibson GR. 1998. Dietary modulation of the human gut microflora using prebiotics. British Journal of Nutrition 80(4): S209-S212.

Gidley MJ, Lillford PJ, Rowlands DW. 1991. Structure and solution properties of tamarind-seed polysaccharide. Carbohydrate Research 214: 299-314.

Goel RK, Tavares IA, Bennett A. 1989. Stimulation of gastric and colonic mucosal eicosonoid synthesis by plantain banana. Journal of Pharmacy and Pharmacology 41: 747-750.

Gopalakrishnan G, Banumathi B, Suresh G. 1997. Evaluation of the antifungal activity of natural xanthones from *Garcinia mangostana* and their synthetic derivatives. Journal of Natural Products 60: 519-524.

Gopalakrishnan C, Shankaranarayanan D, Kameswara L, Nazimudern SK. 1980. Effect of mangostin, a xanthone from *Garcinia mangostana* Linn in immunopathological and inflammatory reactions. Indian Journal of Experimental Biology 18: 843-846.

Gorinstein S, Zemser M, Haruenkit R, Chuthakorn R, Grauer F, et al. 1999. Comparative content of total polyphenols and dietary fiber in tropical fruits and persimmon. Journal of Nutritional Biochemistry 10(6): 367-371.

Grigelmo-Miguel N, Gorinstein S, Martín-Belloso O. 1999. Characterization of peach dietary fiber concentrate as a food ingredient. Food Chemistry 65(2): 175-181.

Herbst ST. 2001. The food lover's companion. Barron's Educational Series, Hauppauge, NY.

Ho C-K, Huang Y-L, Chen C-C. 2002. Garcinone E, a xanthone derivative, has potent cytotoxic effect against hepatocellular carcinoma cell lines. Planta Medica 68: 975-979.

Horowitz RM, Gentili B. 1969. Taste and structure in phenolic glucosides. Journal of Agricultural and Food Chemistry 17: 696-700.

Hossain M, Haque A, Hossain M. 1979. Nutritive value of jackfruit. Bangladesh Journal of Agricultural Sciences 4(1): 9-12.

Hsu FL, Lu FH, Cheng JT. 1994. Influence of acetonylgeraniin, a hydrolyzable tannin from *Euphoria longana,* on orthostatic hypotension in a rat model. Planta Medica 60: 297-300.

Huet, R. 1982. Constituants des agrumes a' effet pharmacody-namique: les citroflavonoid (Constituents of citrus fruits with pharmacodynamic effect: citroflavonoids). Fruits 37: 267- 271.

Hyson D, Studebaker Hallman D, Davis PA, Gershwin ME. 2000. Top of Form Apple juice consumption reduced plasma low-density lipoprotein oxidation in healthy men and women. Journal of Medicinal Food 3(4): 159-166.

Imbabi E, Ibrahim KE, Ahmed BM, Abulfutuh IM, Hulbert P. 1992. Chemical characterization of tamarind bitter principle, tamarindineal [tamarindienal]. Fitoterapia 63(6): 537-538.

Jeon SM, Bok SH, Jang MK, Lee MK, Nam KT, et al. 2001. Antioxidative activity of naringin and lovastatin in high cholesterol-fed rabbits. Life Science 69: 2855-2866.

Joshipura KJ, Ascherio A, Manson JE, Stampfer MJ, Rimm EB, et al. 1999. Fruit and vegetable intake in relation to risk of ischemic stroke. Journal of the American Medical Association 282: 1233-1239.

Joyeux M, Mortier F, Flurentin J. 1995. Screening of antiradical, antilipoperoxidant and hepatotoprotective effects of nine plant extracts used in Caribbean folk medicine. Phytotherapy Research 9: 228-230.

Jung HA, Su BN, Keller WJ, Mehta RG, Kinghorn AD. 2006. Antioxidant xanthones from pericarp of *Garcinia mangostana* (Mangosteen). Journal of Agricultural and Food Chemistry 54: 2077-2082.

Kabir S. 1998. Jacalin: A jackfruit (*Artocarpus heterophyllus*) seed-derived lectin of versatile applications in immunobiological research. Journal of Immunological Methods 212(2): 193-211.

Kabuki T, Nakajima H, Arai M, Ueda S, Kuwabara Y, et al. 2000. Characterization of novel antimicrobial compounds from mango (*Mangifera indica* L.) kernel seeds. Food Chemistry 71(1): 61-66.

Kamei H, Kojima S, Hasegawa M, Umeda T, Terabe K, et al. 1993. Suppressive effect of flavonoid extracts from flower petals on cultured human malignant cells. Tokai Journal of Experimental and Clinical Medicine 164: 829-830.

Kanazawa K, Sakakibara H. 2000. High content of dopamine, a strong antioxidant, in Cavendish banana. Journal of Agricultural and Food Chemistry 48(3): 844-848.

Kato Y, Miyake Y, Yamamoto K, Shimomura Y, Ochi H, et al. 2000. Preparation of a monoclonal antibody to *N*(epsilon)-(hexanonyl)lysine, application to the evaluation of protective effects of flavonoid supplementation against exercise-induced oxidative stress in rat skeletal muscle. Biochemical and Biophysical Research Communications 274: 389-393.

Kaur M, Singh N, Sandhu KS, Guraya HS. 2004. Physicochemical, morphological, thermal and rheological properties of starches separated from kernels of some Indian mango cultivars (*Mangifera indica* L.). Food Chemistry 85: 131-140.

Kharn MA, Nizami SS, Khan MNI, Azeem SW, Ahamed Z. 1994. New triterpenes from *Mangifera indica.* Journal of Natural Products 57: 988-991.

Kim DH, Song MJ, Bae EA, Han MJ. 2000. Inhibitory effect of herbal medicines on rotavirus infectivity. Biological and Pharmaceutical Bulletin 23: 356-358.

Kirtikar KR, Basu BD. 1974. Indian medicinal plants, Vol. 2. M/S Periodical Experts, New Delhi. p. 951.

Knekt P, Jarvinen R, Reunanen A, Maatela J. 1996. Flavonoid intake and coronary mortality in Finland: A cohort study. British Medical Journal 312: 478-481.

Kohno N, Taima M, Sumida T, Azuma Y, Ogawa H, et al. 2001. Inhibitory effect of mandarin juice rich in beta-cryptoxanthin and hesperidin on 4-(methylnitrosamino)-1-(3-pyridyl)-1-butanone-induced pulmonary tumorigenesis in mice. Cancer Letters 174: 141-150.

Komutarin T, Azadi S, Butterworth L, Keil D, Chitsomboon B, et al. 2004. Extract of the seed coat of *Tamarindus indica* inhibits nitric oxide production by murine macrophages *in vitro* and *in vivo*. Food and Chemical Toxicology 42: 649-658.

Kosuge T, Ishida H, Ishii M. 1985. Studies on active substances in the herbs used for oketsu (stagnant blood) in Chinese medicine II: On the coagulation principle in Persicae semen. Chemical and Pharmaceutical Bulletin 33: 1496-1498.

Kratchanova M, Bénémou C, Kratchanov C. 1991. On the pectic substances of mango fruits. Carbohydrate Polymers 15: 271-282.

Kroyer G. 1986. Über die antioxidative Aktivität von Zitrusfruchtschalen (Antioxidant activity of citrus peels). Z Ernaehrungswiss 25: 63-69.

Kurowska EM, Spence JD, Jordan J, Wetmore S, Freeman DJ, et al. 2000. HDL-cholesterol-raising effect of orange juice in subjects with hypercholesterolemia. American Journal of Clinical Nutrition 72: 1095-1100.

Lakshminarayana G, Rao TC, Ramalingaswamy PA. 1983. Varietal variations in content, characteristics and composition of mango seeds and fat. Journal of the American Oil Chemists' Society 60: 88-89.

Lata B, Przeradzka M, Binkowska M. 2005. Great differences in antioxidant properties exist between 56 apple cultivars and vegetation seasons. Journal of Agricultural and Food Chemistry 53(23): 8970-8978.

Latte KP, Kolodziej H. 2000. Antifungal effects of hydrolyzable tannins and related compounds on dermatophytes, mould fungi and yeasts. Z Naturforsch 55c: 467-472.

Lee CH, Jeong TS, Choi YK, Hyun BH, Oh GT, et al. 2001. Anti-atherogenic effect of citrus flavonoids naringin and naringenin, associated with hepatic ACAT and aortic VCAM-1 and MCP-1 in high cholesterol-fed rabbits. Biochemical and Biophysical Research Communications 284: 681-688.

Lee HS, Wicker L. 1991. Quantitative changes in anthocyanin pigments of lychee fruit during refrigerated storage. Food Chemistry 40: 263-270.

Lee JW, Choi JH, Kang SM. 1992. Screening of medicinal plants having hepatoprotective activity: Effects with primary cultured hepatocytes intoxicated using CCl_4 cytotoxicity. Korean Journal of Pharmacognosy 23: 268-275.

Leontowicz M, Gorinstein S, Leontowicz H, Krzeminski R, Lojek A, et al. 2003. Apple and pear peel and pulp and their influence on plasma lipids and antioxidant potentials in rats fed cholesterol-containing diets. Journal of Agricultural and Food Chemistry 51(19): 5780-5785.

Levi F, Pasche C, La Vecchia C, Lucchini F, Franceschi S. 1999. Food groups and colorectal cancer risk. British Journal of Cancer 79: 1283-1287.

Levi F, Pasche C, Lucchini F, Bosetti C, Franceschi S, et al. 2000. Food groups and oesophageal cancer risk in Vaud, Switzerland. European Journal of Cancer Prevention 9: 257-263.

Lewis DA, Shaw GP. 2001. A natural flavonoid and synthetic analogues protect the gastric mucosa from aspirin-induced erosions. Journal of Nutritional Biochemistry 12(2): 95-100.

Li S. 1982. Bencao gangmu. Weisheng Press, Renmin Beijing. p. 1756.

Liu RH, Liu J, Chen B. 2005. Apples prevent mammary tumors in rats. Journal of Agricultural Food Chemistry 53(6): 2341-2343.

Maccarone E, Rapisarda P, Fanella F, Arena E, Mondello L. 1998. Cyanidin-3-(6″-malonyl)-glucoside. An important anthocyanin of blood orange juice. Italian Journal of Food Science 10: 367-372.

Mahabusarakam W, Iriyachitra P, Taylor WC. 1987. Chemical constituents of *Garcinia mangostana*. Journal of Natural Products 50: 474-478.

Mahabusarakum W, Phongpaichit S, Jansakul C, Wiriyachitra P. 1983. Screening of antibacterial activity of chemicals from *Garcinia mangostana*. Songklanakarin Journal of Science and Technology 5: 337-339.

Maiti R, Jana D, Das UK, Ghosh D. 2004. Antidiabetic effect of aqueous extract of seed of *Tamarindus indica* in streptozotocin-induced diabetic rats. Journal of Ethnopharmacology 92: 85-91.

Majumdar PK, Sharma DK. 1985. Mango. In: Bose TK, editor. Fruits of India: Tropical and subtropical. Naya Prokash, Kolkata, India. p. 69-123.

Manthey JA, Grohmann K, Guthrie N. 2001. Biological properties of citrus flavonoids pertaining to cancer and inflammation. Current Medical Chemistry 8: 135-153.

Marini D, Balestrieri F. 1995. Multivariate analysis of flavanone glycosides in citrus juices. Italian Journal of Food Science 3: 255-264.

Martin FW. 1980. Durian and mangosteen. In: Nagy S, Shaw DE, editors. Tropical and subtropical fruits: Composition, properties and uses. AVI, Westport, CT. p. 407-414.

Martinello F, Soares SM, Franco JJ, Santos AC, Sugohara A, et al. 2006. Hypolipemic and antioxidant activities from *Tamarindus indica* L. pulp fruit extract in hypercholesterolemic hamsters. Food and Chemical Toxicology 44(6): 810-818.

Martinez G, Delgado R, Perez G, Garrido G, Nunez Selles AJ, et al. 2000. Evaluation of the *in vitro* antioxidant activity of *Mangifera indica* L. extract (Vimang). Phytotherapy Research 14(6): 424-427.

McCann MJ, Gill CIR, O' Brien G, Rao JR, McRoberts WC, et al. 2007. Anti-cancer properties of phenolics from apple waste on colon carcinogenesis in vitro. Food and Chemical Toxicology 45(7): 1224-1230.

McCullough ML, Robertson AS, Jacobs EJ, Chao A, Calle EE, et al. 2001. A prospective study of diet and stomach cancer mortality in United States men and women. Cancer Epidemiology Biomarkers and Prevention 10: 1201-1205.

Mercadante AZ, Rodriguez-Amaya DB. 1998. Effects of ripening, cultivar differences and processing on the carotenoid composition of mango. Journal of Agricultural and Food Chemistry 46: 128-130.

Miller NJ, Sampson J, Candeias LP, Bramley PM, Rice-Evans CA. 1996. Antioxidant activities of carotenes and xanthophylls. FEBS Letters 384: 240-242.

Mishra AK, Dubey NK. 1990. Fungitoxic properties of *Prunus persica* oil. Hindustan Antibiotic Bulletin 32: 91-93.

Misra P, Pal NL, Guru PY, Katiyar JC, Tandon JS. 1991. Antimalarial activity of traditional plants against erythrocytic stages of *Plasmodium berghei*. International Journal of Pharmacognosy 29: 19-23.

Miyake Y, Shimoi K, Kumazawa S, Yamamoto K, Kinae N, et al. 2000. Identification and antioxidant activity of flavonoid metabolites in plasma and urine of eriocitrin-treated rats. Journal of Agricultural and Food Chemistry 48: 3217-3224.

Moharram YG, Moustafa AM. 1982. Utilisation of mango seed kernel (*Mangifera indica*) as a source of oil. Food Chemistry 8: 269-276.

Moongkarndi P, Kosem N, Kaslungka S, Luanratana O, Pongpan N, et al. 2004. Antiproliferation, antioxidation and induction of apoptosis by *Garcinia mangostana* (mangosteen) on SKBR3 human breast cancer cell line. Journal of Ethnopharmacology 90(1): 161-166.

Mortin JF. 1987. Longan. In: Julia F, editor. Fruits of warm climates. Morton, Miami, FL. p. 259-262.

Mouly PP, Arzouyan CR, Gaydou EM, Estienne JM. 1994. Differentiation of citrus juices by factorial discriminant analysis using liquid chromatography of flavanone glycosides. Journal of Agricultural and Food Chemistry 42: 70-79.

Murakami A, Nakamura Y, Ohto Y, Yano M, Koshiba T, et al. 2000. Suppressive effects of citrus fruits on free radical generation and nobiletin, an anti-inflammatory polymethoxyflavonoid. Biofactors 12(1-4): 187-192.

Nadkarni KM. 1976. Indian materia medica. 3rd ed., Popular Prakashan, Bombay. p. 1036-1037.

Nagamoto N, Naguchi H, Nanba K, Nakamura H, Mizuno M. 1988. Active components having anti-inflammatory and analgesic activities from *Armeniacae semen*, *Pruni japonicae semen* and almond seeds. Shoyakugaku Zasshi 42(1): 81-88.

Nakasone HY, Paull RE. 1998. Tropical fruits. CAB International, Wallingford, UK.

Nilar, Nguyen L-HD, Venkatraman G, Sim K-Y, Harrison LJ. 2005. Xanthones and benzophenones from *Garcinia griffithii* and *Garcinia mangostana*. Phytochemistry 66: 1718-1723.

Notka F, Meier GR, Wagner R. 2003. Inhibition of wild-type human immunodeficiency virus and reverse transcriptase inhibitor-resistant variants by *Phyllanthus amarus*. Antiviral Research 58: 175-186.

Okabe S, Suganuma M, Imayoshi Y, Taniguchi S, Yoshida T, Fujiki H. 2001. New TNF-α releasing inhibitors, geraniin and corilagin, in leaves of *Acer nikoense*, Megusurino-ki. Biological and Pharmaceutical Bulletin 24: 1145-1148.

Okuyama E, Ebihara H, Takeuchi H, Yamazaki M. 1999. Adenosine, the anxiolytic-like principle of the arillus of *Euphoria longana*. Planta Medica 65: 115-119.

Omoregie E. 2001. Utilization and nutrient digestibility of mango seeds and palm kernel meal by juvenile *Labeo senegalensis* (Antheriniformes: Cyprinidae). Aquaculture Research 32: 681-687.

Ooghe WC, Detavernier CM. 1997. Detection of the addition of *Citrus reticulata* and hybrids to *Citrus sinensis* by flavonoids. Journal of Agricultural and Food Chemistry 45: 1633-1637.

Palli D, Russo A, Ottini L, Masala G, Saieva C, et al. 2001. Red meat, family history, and increased risk of gastric cancer with microsatellite instability. Cancer Research 61: 5415-5419.

Pandey M, Shukla VK. 2002. Diet and gallbladder cancer: A case-control study. European Journal of Cancer Prevention 11(4): 365-368.

Pari L, Maheswari JU. 1999. Hypoglycaemic effect of *Musa sapientum* L. in alloxan-induced diabetic rats. Journal of Ethnopharmacology 68(1-3): 321-325.

Pari L, Maheswari JU. 2000. Antihyperglycaemic activity of *Musa sapientum* flowers: Effect on lipid peroxidation in alloxan diabetic rats. Phytotherapy Research 14(2): 136-138.

Pedroza-Islas R, Aguilar-Esperanza E, Vernon-Carter EJ. 1994. Obtaining pectins from solid wastes derived from mango (*Mangifera indica*) processing. AIChE Symposium Series 300: 36-41.

Pereira RB, Mancini-Filho J. 1994. Avaliaçao da actividade antioxidante em sementes de frutas citricas [Antioxidant activity of citrus seeds]. Ciência e Tecnologia de Alimentos 14: 160-167.

Piskur MM, Higgins JW. 1949. Antioxidants for fats. U.S. Patent 2461080.

Pott I, Marx M, Neidhart S, Mühlbauer W, Carle R. 2003. Quantitative determination of β-carotene stereoisomers in fresh, dried and solar-dried mangoes (*Mangifera indica* L.). Journal of Agricultural and Food Chemistry 51: 4527-4531.

Puravankara D, Boghra V, Sharma RS. 2000. Effect of antioxidant principles isolated from mango (*Mangifera indica* L.) seed kernels on oxidative stability of buffalo ghee (butter-fat). Journal of the Science of Food and Agriculture 80: 522-526.

Ramos A, Visozo A, Piloto J, Garcia A, Rodriguez CA, et al. 2003. Screening of antimutagenicity via antioxidant activity in Cuban medicinal plants. Journal of Ethnopharmacology 87: 241-246.

Ramulu P, Rao PU. 2003. Total, insoluble and soluble dietary fiber contents of Indian fruits. Journal of Food Composition and Analysis 16(6): 677-685.

Rao AV, Agarwal S. 1999. Role of lycopene as antioxidant carotenoid in the prevention of chronic disease: A review. Nutrition Research 19: 305-323.

Rao YS, Mathew KM, Potty SN. 1999. Tamarind (*Tamarindus indica* L.) research: A review. Indian Journal of Arecanut, Spices and Medicinal Plants 1(4): 127-145.

Rapisarda P, Tomaino A, Lo Cascio R, Bonina F, De Pasquale A, et al. 1999. Antioxidant effectiveness as influenced by phenolic content of fresh orange juices. Journal of Agricultural and Food Chemistry 47: 4718-4723.

Rastogi RP, Mehrotra BN. 1995. Compendium of Indian medicinal plants. CDRI, Lucknow and Publications and Information Directorate, New Delhi.

Ravindran V, Sivakanesan R. 1996. The nutritive value of mango seed kernels for starting chicks. Journal of the Science of Food and Agriculture 71: 245-250.

Remani P, Bhattathiri VN, Bindu L, Chandralekha B, Vijayakumar T, et al. 1997. Correlation of lectin binding with lymph node metastasis in oral cancers. Oral Oncology 33(1): 19-22.

Rice-Evans CA. 1999. Screening of phenolics and flavonoids for antioxidant activity. In: Packer L, Hiramatsu M, Yoshikawa T, editors. Antioxidant food supplements in human health. Academic Press, San Diego, CA. p. 239-253.

Rice-Evans CA, Miller NJ, Paganga G. 1997. Antioxidant properties of phenolic compounds. Trends in Plant Science 2: 152-159.

Rimbau V, Cerdan C, Vila R, Iglesias J. 1999. Antiinflammatory activity of some extracts from plants used in the traditional medicine of north-African countries (II). Phytotherapy Research 13(2): 128-132.

Ruiz D, Egea J, Tomás-Barberán FA, Gil MI. 2005. Carotenoids from new apricot (*Prunus armeniaca* L.) varieties and their relationship with flesh and skin color. Journal of Agricultural and Food Chemistry 53(16): 6368-6374.

Saleh NA, El-Ansari MA. 1975. Polyphenolics of twenty local varieties of *Mangifera indica*. Planta Medica 28: 124-130.

Sanchez GM, Re L, Giuliani A, Nunez-Selles AJ, Davison GP, et al. 2000. Protective effects of *Mangifera indica* L. extract, mangiferin and selected antioxidants against TPA-induced biomolecules oxidation and peritoneal macrophage activation in mice. Pharmacology Research 42(6): 565-573.

Sankaranarayanan R, Sekar K, Banerjee R, Sharma V, Surolia A, et al. 1996. A novel mode of carbohydrate recognition in jacalin, a Moraceae plant lectin with a beta-prism fold. Nature Structural Biology 3(7): 596-603.

Sarni-Manchado P, Le Roux E, Le Guerneve C, Lozano Y, Cheynier V. 2000. Phenolic composition of litchi fruit pericarp. Journal of Agricultural and Food Chemistry 48: 5995-6002.

Satoru K, Yasuhiko T, Eriko K, Kazunori O, Masamichi Y. 1999. Antiproliferative effects of the readily extractable fractions prepared from various citrus juices on several cancer cell lines. Journal of Agricultural and Food Chemistry 47(7): 2509-2512.

Satyavati GV, Raina MK, Sharma M. 1976. Medicinal plants of India. Cambridge Printing Works, New Delhi, India.

Sawamura M, Kuriyama T, Li Z. 1988. Rind spot, antioxidative activity and tocopherols in the flavedo of citrus fruits. Journal of Horticultural Science and Biotechnology 63: 717-721.

Scartezzini P, Speroni E. 2000. Review on some plants of Indian traditional medicine with antioxidant activity. Journal of Ethnopharmacology 71(1-2): 23-43.

Schieber A, Berardini N, Carle R. 2003. Identification of flavonol and xanthone glycosides from mango (*Mangifera indica* L. cv. 'Tommy Atkins') peels by high-performance liquid chromatography-electrospray ionization mass spectrometry. Journal of Agricultural and Food Chemistry 51: 5006-5011.

Selles NAJ, Castro HTV, Aguero-Aguero J, Gonzalez J, Nadeo F, et al. 2002. Isolation and quantitative analysis of phenolic antioxidants, free sugars, and polyols from mango (*Mangifera indica* L.) stem bark aqueous decoction used in Cuba as a nutritional supplement. Journal of Agricultural and Food Chemistry 50: 762-766.

Sergio W. 1998. A natural food, the Malabar tamarind, may be effective in the treatment of obesity. Medical Hypotheses 27(1): 39-40.

Sesso HD, Buring JE, Norkus EP, Gaziano JM. 2004. Plasma lycopene, other carotenoids and retinol and the risk of cardiovascular disease in women. American Journal of Clinical Nutrition 79: 47-53.

Setiawan B, Sulaeman A, Giraud DW, Driskell JA. 2001. Carotenoid content of selected Indonesian fruits. Journal of Food Composition and Analysis 14(2): 169-176.

Shiota S, Shimizu M, Sugiyama J, Morita Y, Mizushima T, et al. 2004. Mechanisms of action of corilagin and tellimagrandin I that remarkably potentiate the activity of beta-lactams against methicillin-resistant *Staphylococcus aureus*. Microbiology and Immunology 48: 67-73.

Singh UP, Singh DP, Singh M, Maurya S, Srivastava JS, et al. 2004. Characterization of phenolic compounds in some Indian mango cultivars. International Journal of Food Science and Nutrition 55: 163-169.

Solis CS. 1969. Antibacterial and antibiotic properties of the Leguminosae. Acta Manilana Series A 4: 52-109.

Sood AK, Kapil U. 1990. Knowledge and practices among rural mothers in Haryana about childhood diarrhea. Indian Journal of Pediatrics 57: 563-566.

Sreelekha TT, Vijayakumar T, Ankanthil R, Vijayan KK, Nair MK. 1993. Immunomodulatory effects of a polysaccharide from *Tamarindus indica*. Anti-Cancer Drugs 4(2): 209-212.

Srirangarajan AN, Shrikhande AJ. 1976. Mango peel waste as a source of pectin. Current Science 45: 620-621.

Srirangarajan AN, Shrikhande AJ. 1977. Characterization of mango peel pectin. Journal of Food Science 42: 279-280.

Srirangarajan AN, Shrikhande AJ. 1979. Comparative aspects of pectin extracted from the peels of different varieties of mango. Journal of Food Technology 14: 539-541.

Sudhakar DV, Maini SB. 2000. Isolation and characterization of mango peel pectins. Journal of Food Processing and Preservation 24: 209-227.

Sujathan K, Kannan S, Remani P, Pillai KR, Chandralekha B, et al. 1996. Differential expression of jackfruit-lectin-specific glycoconjugates in metastatic adenocarcinoma and reactive mesothelial cells: A diagnostic aid in effusion cytology. Journal of Cancer Research and Clinical Oncology 122(7): 433-436.

Suksamrarn S, Suwannapoch N, Phakhodee W, Thanuhiranlert J, Ratananukul P, et al. 2003. Antimycobacterial activity of prenylated xanthones from the fruits of *Garcinia mangostana*. Chemical and Pharmaceutical Bulletin 51: 857-859.

Suksamrarn S, Suwannapoch N, Ratananukul P, Aroonlerk N, Suksamrarn A. 2002. Xanthones from the green fruit hulls of *Garcinia mangostana*. Journal of Natural Products 6: 761-763.

Sun J, Chu Y-F, Wu X, Liu RH. 2002. Antioxidant and antiproliferative activities of common fruits. Journal of Agricultural and Food Chemistry 50(25): 7449-7454.

Tandon DK, Garg N. 1999. Mango waste: A potential source of pectin, fibre and starch. Indian Journal of Environmental Protection 19: 924-927.

Tandon DK, Kalra SK, Singh BP, Garg N. 1991. Characterization of pectin from mango fruit waste. Indian Food Packer 45: 9-12.

Tian Q, Miller EG, Ahmad H, Tang L, Patil BS. 2001. Differential inhibition of human cancer cell proliferation by citrus limonoids. Nutrition and Cancer 40: 180-184.

Ting, SV, Newhall, WF. 1965. The occurrence of a natural antioxidant in citrus fruit. Food Research 30: 57-63.

Tona L, Kambu K, Ngimbi N, Cimanga K, Vlietinck AJ. 1998. Antiamoebic and phytochemical screening of some Congolese medicinal plants. Journal of Ethnopharmacology 61(1): 57-65.

Tsao R, Yang R, Young JC, Zhu H. 2003. Polyphenolic profiles in eight apple cultivars using high-performance liquid chromatography (HPLC). Journal of Agricultural and Food Chemistry 51: 6347-6353.

Tsuda T, Fukaya Y, Ohshima K, Yamamoto A, Kawakishi S, et al. 1995. Antioxidative activity of tamarind extract prepared from the seed coat. Nippon Shokuhin Kagaku Kogaku Kaishi 42(6): 430-435.

Tsuda T, Watanabe M, Ohshima K, Yamamoto A, Kawakishi S, et al. 1994. Antioxidative components isolated from the seed of tamarind (*Tamarindus indica* L.). Journal of Agricultural and Food Chemistry 42(12): 2671-2674.

Vanamala J, Leonardi T, Murphy ME, Taddeo SS, Patil B, et al. 2004. Natural and irradiated grapefruit pulp and their functional compounds suppress aberrant crypt foci and colonocyte proliferation. Experimental Biology, April 17-21, 2004, Washington, DC.

Vijayan KK, Remani P, Beevi VM, Ankathil R, Vijayakumar T, et al. 1987. Tissue binding patterns of lectins in premalignant and malignant lesions of the oral cavity. Journal of Experimental Pathology 3(3): 295-304.

Wall MM. 2006a. Ascorbic acid, vitamin A, and mineral composition of banana (*Musa* sp.) and papaya (*Carica papaya*) cultivars grown in Hawaii. Journal of Food Composition and Analysis 19(5): 434-445.

Wall MM. 2006b. Ascorbic acid and mineral composition of longan (*Dimocarpus longan*), lychee (*Litchi chinensis*) and rambutan (*Nephelium lappaceum*) cultivars grown in Hawaii. Journal of Food Composition and Analysis19(6-7): 655-663.

Wang H, Nair MG, Strasburg GM, et al. 1999. Antioxidant and antiinflammatory activities of anthocyanins and their aglycon, cyanidin, from tart cherries. Journal of Natural Products 62(2): 294-296.

Wang X, Wei Y, Yuan S, Liu G, Lu Y, et al. 2005. Potential anticancer activity of litchi fruit pericarp extract against hepatocellular carcinoma *in vitro* and *in vivo*. Cancer Letters 239(1): 144-150.

Wang X, Yuan S, Wang J, Lin P, Liu G, et al. 2006. Anticancer activity of litchi fruit pericarp extract against human breast cancer *in vitro* and *in vivo*. Toxicology and Applied Pharmacology 215(2): 168-178.

Weecharangsan W, Opanasopit P, Sukma M, Ngawhirunpat T, Sotanaphun U, et al. 2006. Antioxidative and neuroprotective activities of extracts from the fruit hull of mangosteen (*Garcinia mangostana* Linn.). Medical Principles and Practice: International Journal of Kuwait University 15(4): 281-287.

Wilcox LJ, Borradaile NM, de Dreu LE, Huff MW. 2001. Secretion of hepatocyte apoB is inhibited by the flavonoids, naringenin and hesperetin, via reduced activity and expression of ACAT2 and MTP. Journal of Lipid Research 45: 725-734.

Willett WC. 1995. Diet, nutrition, and avoidable cancer. Environmental Health Perspectives 103: 165-170.

Williams NS, Harris ND. 1983. Antioxidant activity in dried orange. Journal of Food Science 48: 644-645.

Williams P, Ongsakul M, Proudfoot J, Croft K, Bellin L. 1995. Mangostin inhibits the oxidative modification of human low density lipoprotein. Free Radical Research 23: 175-184.

XanGo. 2006. Supplement Facts. Available at: http://www.xango.net/drink/bottle .html. Accessed April 6, 2006.

Xing N, Chen Y, Mitchell SH, Young CYF. 2001. Quercetin inhibits the expression and function of the androgen receptor in LNCaP prostate cancer cells. Carcinogenesis 22: 409-414.

Yi H-Z, Yang S-Y, Li K. 2005. Analysis of nutrient components in the fruits of sweet Thai tamarind (*Tamarindus indica* L.) cultivars. Journal of Plant Resources and Environment 14(3): 57-58.

Yoshikawa M, Harada E, Miki A, Tsukamoto K, Liang SQ, et al. 1994. Antioxidant constituents from the fruit hulls of mangosteen (*Garcinia mangostana*) originating in Vietnam. Yakugaku Zassh 114: 129-133.

Zhang Z, Pang X, Yang C, Ji Z, Jiang Y. 2004. Purification and structural analysis of anthocyanins from litchi pericarp. Food Chemistry 84: 601-604.

Zhang DL, Quantick PC, Grigor JM. 2000a. Changes in phenolic compounds in litchi (*Litchi chinensis* Sonn.) fruit during postharvest storage. Postharvest Biology and Technology 19: 165-172.
Zhang M, Zhang JP, Ji HT, Wang JS, Qian DH. 2000b. Effect of six flavonoids on proliferation of hepatic stellate cells *in vitro*. Acta Pharmceutica Sinica 21: 253-256.
Zheng G, Yi Z, Zhang J, Zhong D. 2003. Studies on the antioxidative effect of extract from mature and premature litchi pericarp. Natural Products Research and Development 15: 341-344.
Zhu XM, Song JX, Huang ZZ, Wu YM, Yu MJ. 1993. Antiviral activity of mangiferin against herpes simplex virus type 2 *in vitro*. Chung-Kuo-Yao-Li-Hsueh-Pao 15: 452-454.

Chapter 4

Fats and Oils

Fats are essential nutrients in the human diet and are normally included in the diet in the form of cooking fat and oils from plant or animal sources. Fats supply energy and aid in nutrient transport. However, obesity worldwide (Bell et al. 2001; Flegal et al. 2002) and chronic health issues associated with a high-fat diet, including cardiovascular disease, cancer, and diabetes, have turned the attention of researchers to the nature of dietary fat and sources of fat in the human diet. Dietary fat and fatty acids are considered important risk factors in carcinogenesis and cardiovascular diseases. Common plant sources of fats are oilseed crops, including sesame, peanut, coconut, canola (or rapeseed), soybean, palm, and blends of one or more of these oils commonly termed *vegetable oil;* these oils are used in cooking and preparation of food. An oil base in cooking foods also serves as a hydrophobic medium to enhance the bioavailability of hydrophobic phytochemicals and antioxidant vitamins. Vegetable oils are produced mainly from oil seeds (e.g., rapeseed and sunflower seed) as well as from the flesh of some fruits (e.g., olives). Vegetable oils are often a rich source of unsaturated fatty acids, although some saturates are also present. Vegetable oils usually contain natural antioxidants, such as vitamin E, which help the oils resist the rancidity that occurs when oxidation takes place. Sometimes a blend of oils is used in a product to take advantage of the desirable characteristics of the different types of oils. Some oils are rich in omega-6 fatty acids (e.g., sunflower and corn oils), while others provide omega-3 fatty acids (e.g., rapeseed and soy oils). Other oils are rich in monounsaturated fatty acids (e.g., olive, rapeseed, and groundnut oils).

Asian Crops and Human Dietetics

doi:10.1300/5504_04

SESAME

According to some archaeological findings, sesame (*Sesamum indicum;* Pedaliaceae) is the most ancient oilseed (Bedigian and Harlan 1986; Nayar 1984). Its cultivation goes back to 2130 BC (Weiss 1983). Also known as gingili, sesamum, gingelly, beniseed, sim-sim, and till, sesame is an important oilseed that originated in tropical Africa and India. In some parts of Africa and Asia it is called the queen of oilseed crops because of its high yield, usage, and economic returns. India and China are the world's largest producers of sesame, followed by Burma, Sudan, Mexico, Nigeria, Venezuela, Turkey, Uganda, Ethiopia, Japan, and the southwestern United States.

Sesame seeds have a delicious taste and are used to flavor bread, biscuits, candy, and other desserts and delicacies. It is often sprinkled over Japanese and Korean dishes as a garnish or used as a main ingredient in some rice and curried dishes. Tahini, or sesame butter, is a paste made from ground dried sesame seeds used to thicken and flavor sauces and gravies in Middle East and Arab countries. It is also an important ingredient in hummus, a bread spread popular in Israel, Egypt, Lebanon, and other Arab nations, and chutneys and pickles of Indian cuisine. The dried seeds are often toasted to increase the flavor. The seeds contain 50% to 60% edible oil, which has a unique composition of monounsaturated and polyunsaturated fatty acids. The oxidative stability of sesame oil is superior to that of other vegetable oils, although it contains nearly 85% unsaturated fatty acids (Sonntag 1981), because of its excellent antioxidant potential. Sesame oil is a major culinary and massage oil because of its attributed medicinal properties. Ayurveda considers sesame oil to be sweet and act as an antitoxin and anti-inflammatory. Ayurvedic practice favors sesame oil for massage of children and adults because it promotes good muscle development and healthy hair and skin. The Unani medical system also uses sesame as a mild laxative and in hair cosmetic preparations, liniments, and poultices.

Sesame is an erect annual plant that grows to a height of about 0.6 m (2 ft), depending on the variety and the growing conditions. Some varieties produce numerous branches, while others do not produce any branches. The leaves are oblong, pubescent, and rough to touch; flowers are bell shaped, white, pale pink, or purplish; and the fruit is a capsule 5 cm to 10 cm (2-4 in) long containing numerous seeds. Sesame seeds are small and vary from off-white or cream to black.

Sesame is recognized as an important source of protein, calcium, phosphorus, and iron (Alobo 2001; Al-Yemeni et al. 2000). It is used in the manufacture of health foods (Miyake 1999). Some people are allergic to sesame seeds because they are unable to digest the complex and unique protein present in sesame seeds (Pastorello et al. 2001). The attributed benefits of sesame in skin and hair care in Asian traditional medicine and the demonstrated positive results have triggered the use of sesame as an ingredient in numerous patented cosmetics for skin and hair care (Fu 1996; Kurita and Shimogaki 2001; Kuriyama et al. 1998; Simon and Chevalier 2000; Yamaha et al. 2000).

The antiaging properties of sesame oil are proposed to result from enhancement and stimulation of vitamin E activity in cells and the synergistic effect of its lignans and tocopherols (Kuno et al. 1999; Yamashita 1997). Other medicinal properties of sesame seeds and oil include hypolipidemic, hypoglycemic (Anila and Vijayalakshmi 2000; Sirato-Yasumoto et al. 2001), and antioxidant (Osawa 1999; Potterat 1997; Yun 1999), and it is used as an antiviral against hepatitis B virus (Hussein et al. 2000). Sesame oil has been suggested to decrease blood pressure (Sankar et al. 2005), increase the liver's lipolytic activity (Sirato-Yasumoto et al. 2001), lower blood cholesterol levels (Sankar et al. 2005), and improve antioxidant capacity in hypercholesterolemic patients (Chen et al. 2005). In 2005 global sesame production was 3,321,458 Mt and the area cultivated was 7,561,200 ha (FAOSTAT 2006).

COCONUT

The coconut (*Cocos nucifera;* Arecaceae) was first mentioned in AD 545 by the Egyptian monk Cosmos Indicopleustes who visited western India and Sri Lanka. In his *Topographia Christiana,* Cosmos described coconut as the "great nut of India." The planting of coconuts in Sri Lanka is described in the Mahavasma in AD 589. In Marco Polo's writings from 1280, he described coconut as *nux indica,* the Indian nut, growing in Sumatra and in India (Madras and Malabar). Coconut is a major commercial crop in the Philippines, Indonesia, and India and on many islands in the Pacific Ocean, the Indian Ocean, and the Caribbean. Coconut oil is the major oil applied for hair styling, and it is also regularly used as a bathing oil in many parts of India. It is a major cooking oil all over India and in several other countries of Southeast Asia.

Ayurvedic medicine uses coconut oil for body massage and for treating skin inflammation, earache, and mouth sores. Coconut roots are believed to be antiblennorrhagic, antibronchitic, antipyretic, and antigingivitic. The roots are used to make toothbrushes, and an extract of the root is used for mouthwash and medicine for dysentery. According to Hartwell (1967-1971), coconuts are folk remedies for treating tumors. Other medicinal uses of coconut include as an anthelmintic, antiseptic, laxative, aphrodisiac, and astringent and as a bactericidal, depurative, diuretic, purgative, antipyretic, and stomachic. Coconut has also been a folk remedy for abscesses, amenorrhea, asthma, blennorrhagia, bronchitis, bruises, burns, colds, constipation, cough, debility, dropsy, dysentery, dysmenorrhea, earache, erysipelas, fever, flu, gingivitis, gonorrhea, hematemesis, hemoptysis, jaundice, menorrhagia, nausea, rash, scabies, scurvy, sore throat, stomach, swelling, syphilis, toothache, tuberculosis, tumors, typhoid, venereal disease, and wounds (Duke and Wain 1981).

The tropical palm tree is native to India and widely distributed throughout Asia, Africa, Latin America, New Zealand, and the Pacific region.

The tree has a woody trunk 30 to 45 cm in diameter rising from a swollen base and is surrounded by a mass of roots. It bears a crown of large pinnate leaves 1.8 to 20 m (6 to 20 ft) long, which are folded like a fan with long, narrow, and tapering leaflets. The inflorescence is a spathe borne in the leaf axil, enclosed in a long spadix, pale yellow or orange and branched. Male flowers of the monoecious coconut tree are at the end of the inflorescence; female flowers are larger than the male, light yellow, and are found toward the base of the inflorescence. The fruits are ovoid, three-angled, and monocotyledon. The thick fibrous husk (exocarp) encloses a hard, bony shell (endocarp). The edible white coconut meat adheres to the shell's inside wall.

The white, fleshy part of immature nuts (approximately four to five months old) is edible, tender, and sweet, and the cavity is filled with a sweet juice called coconut milk. The white meat of the nut can be eaten raw or shredded and dried and is used in many of the world's cuisines. The dried meat of the kernels, known as copra, is crushed to produce coconut oil. The oil can be used for culinary purposes and to produce cosmetics and soap. Grated coconut is also used in coffee and tea as a substitute for cream. A fresh beverage is obtained by incision of the flower clusters of coconut that can be boiled down to a syrup called coconut molasses and then crystallized into dark brown sugar. The liquid also ferments quickly into an alcoholic beverage called *toddy* in India and Sri Lanka, *tuba* in the Philippines and Mexico, and *tuwak* in Indonesia. In east Asia toddy is distilled to make *arrack,* a liquor.

Allowed to germinate, the nut becomes a spongy mass called bread and is eaten raw or toasted in the shell over fire. The outer shell is hard and fine grained and is widely used for carving objects such as drinking cups, dippers, scoops, and smoking pipe bowls and collecting rubber latex. Coconut shell flour is used as a filler in the plastics industry. Fresh, dried, desiccated, and shredded coconut is used in cakes, pies, candies, curries, and sweets.

Coconut oil is used to make soap, shampoo, shaving cream, dentifrices, lotions, lubricants, hydraulic fluid, paint, synthetic rubber, plastics, and margarine and in ice cream. In India coconut oil is used to make a vegetarian ghee (vanaspathi) and as a hair oil, body oil, and burning oil for lamps. Cake residue after the oil has been extracted is rich in protein and carbohydrate and is used as cattle fodder. The high levels of saturated fatty acids in coconut and coconut oil have led to concern over the atherogenecity of coconut and palm oils in food products. However, this perceived negative health effect of coconut consumption has been questioned and some believe that studies demonstrating a hypercholesterolemic effect have been confounded by first developing an essential fatty acid deficiency in the subjects (Elson 1992) and that the adverse health effects of coconut and lauric oils in the diet have been exaggerated (Gurr 1994). Other studies have shown no definite link between coconut and coconut oil and coronary heart disease (Kumar 1997) or cholesterol levels (Cox et al. 1998). Recent research indicates that inclusion of coconut fiber in the diet results in significant hypoglycemic action (Sindurani and Rajamohan 2000) and possible prevention of colon cancer (Manoj et al. 2001; Nalini et al. 1997). In 2005 global sesame production was 55,014,524 Mt and the area cultivated was 10,825,856 ha (FAOSTAT 2006).

FLAX

Also known as linseed, flax (*Linum usitatissimum;* Linaceae) is a native of southwest Asia and southeastern Europe. Flaxseed and flaxseed oil have been used in food in Europe, Africa, and Asia since 5,000 to 8,000 years ago. The fiber of flax stem has been used for linen cloth and many other uses. *Linum* includes approximately 200 species (Carter 1993).

Flax is an erect annual growing to 1.2 m (3.9 ft) with slender stems. The leaves are yellow green, slender, lanceolate, and 2 to 4 cm long and 3 mm broad; the five-petal flowers are a pure pale blue and 1.5 to 2.5 cm in diameter. The fruit is a round, dry capsule 5 to 9 mm in diameter, containing several glossy brown seeds. Flaxseeds, containing about 40% oil, have long been used in human and animal diets and in industry as a source of oil and as the basic component of or an additive to various paints or polymers. Flax oil is the richest plant source of linoleic and linolenic polyunsaturated fatty acids, which humans cannot synthesize and must ingest.

Several authors have published details of flaxseed composition, among them Hettiarachchy et al. (1990), who determined the average composition of 11 flax cultivars, shown in Table 4.1. Flaxseed is also high in dietary fiber and lignans, which are popular as anticarcinogenic compounds (Lay et al. 1989). Ground or whole flaxseeds added to baked products such as bread, waffles, and pancakes impart a nutty flavor. Flaxseeds are used in breakfast cereals and breakfast drinks; salad dressings are made with cold-pressed flaxseed oil; and flaxseed is used in salad toppings, biscuits, meat extenders, crackers, soups, bagels, fiber bars, and cakes. Flaxseed is also ground and used as flour in breads. Interest has grown recently in flax's probiotic properties, high-fiber content, and beneficial effects on coronary heart disease, cancers, and neurological and hormonal disorders (Huang and Milles 1996; Huang and Ziboh 2001; Simopoulos 2002). Other studies have documented the benefit of dietary flax in reducing the symptoms of attention-deficit/hyperactivity disorder in children (Joshi et al. 2006).

TABLE 4.1. Composition of Whole Flaxseed (Average of 11 Cultivars)

Component	Content (%)
Moisture	7.1-8.3
Lipids (DW basis)	31.9-37.8
Protein	26.9-31.6
Total dietary fiber	36.7-46.8
Insoluble	30
Soluble	10
Fatty acid composition	
Palmitic	4.6-6.3
Stearic	3.3-6.1
Oleic	19.3-29.4
Linoleic	14.0-18.2
Linolenic	44.6-51.5

Source: Hettiarachchy et al. 1990.

Note: DW = dry weight.

RAPESEED

Rapeseed, or canola (*Brassica napus* L., *B. rapa* L.; Brassicaceae), has its origin in the Himalayan region, where it was originally used in oil lamps. Rapeseed plants grow to 1 to 1.5 m (3-5 ft). Its leaves are alternate with a waxy coating; flowers are yellow and small (1 cm wide) and borne in clusters of four petals, characteristic of the crucifers. The fruits are narrow pods, usually ascending, and 3 to 5 cm long. Rapeseed oil's nutritional properties are responsible for its domination of the salad-oil market: of the commercially available edible oils, rapeseed contains the lowest level of saturated fats (6%), the second highest level of monounsaturated fats (58%), and the highest level of the essential fatty acid linoleic acid (10%). Rapeseed oil in the diet lowered cholesterol levels and oxidation susceptibility of lipids (Nielsen et al. 2002), and it reduced cholesterol and triglyceride levels (Gulesserian and Widhalm 2002).

REFERENCES

Alobo AP. 2001. Effect of sesame seed flour on millet biscuit characteristics. Plant Foods for Human Nutrition 56(2): 195-202.

Al-Yemeni MN, Hussain MA, Basahy AY. 2000. Mineral composition of some sesame seeds (*Sesamum indicum* L.) grown in the Gizan area of Saudi Arabia. Phyton 67: 121-125.

Anila L, Vijayalakshmi NR. 2000. Beneficial effects of flavonoids from *Sesamum indicum*, *Emblica officinalis*, and *Momordica charantia*. Phytotherapy Research 14(8): 592-595.

Bedigian D, Harlan JR. 1986. Evidence for cultivation of sesame in the ancient world. Economic Botany 40: 137-154.

Bell AC, Ge K, Popkin BM. 2001. Weight gain and its predictors in Chinese adults. International Journal of Obesity and Related Metabolic Disorders 25(7): 1079-1086.

Carter JF 1993. Potential of flaxseed and flaxseed oil in baked goods and other products in human nutrition. Cereal Foods World 38: 753-759.

Chen PR, Chien KL, Su TC, Chang CJ, Liu TL, et al. 2005. Dietary sesame reduces serum cholesterol and enhances antioxidant capacity in hypercholesterolemia. Nutrition Research 25(6): 559-567.

Cox C, Sutherland W, Mann J, De Jong S, Chisholm A, et al. 1998. Effects of dietary coconut oil, butter and safflower oil on plasma lipids, lipoproteins and lathosterol levels. Central European Journal of Public Health 52(9): 650-654.

Duke JA, Wain KK. 1981. Medicinal plants of the world. A computer index with more than 85,000 entries. 3 vols. Food and Agricultural Organization, Rome.

Elson CE. 1992. Tropical oils: Nutritional and scientific issues. Critical Reviews in Food Science and Nutrition 31(1-2): 79-102.

FAOSTAT (Statistics Division of the Food and Agriculture Organization of the United Nations). 2006. http://faostat.fao.org. Accessed February 2006.

Flegal KM, Carroll MD, Ogden CL, Johnson CL. 2002. Prevalence and trends in obesity among US adults, 1999-2000. Journal of the American Medical Association 288(14): 1723-1727.

Fu J. 1996. Compositions containing Chinese medicines for promoting hair growth. Faming Zhuanli Shenqing Gongkai Shuomingshu. p. 13.

Gulesserian T, Widhalm K. 2002. Effect of a rapeseed oil substituting diet on serum lipids and lipoproteins in children and adolescents with familial hypercholesterolemia. Journal of the American College of Nutrition 21(2): 103-108.

Gurr MI. 1994. Nutritional aspects of lauric oils. In: Applewhite TH, editors, Proceedings of the world conference on lauric oils: Sources, processing and applications. AOCS Press, Champaign, IL. p. 104-109.

Hartwell JL. 1967-1971. Plants used against cancer. A survey. Lloydia 1971:30-34.

Hettiarachchy N, Hareland G, Ostenson A, Blader-Shank G. 1990. Composition of eleven flaxseed varieties grown in North Dakota. Proceedings of 53rd Flax Institute of the United States, Fargo, ND. p. 36-40.

Huang Y-S, Milles DE. 1996. Gamma-linolenic acid: Metabolism and its roles in nutrition and medicine. AOCS Press, Champaign, IL.

Huang Y-S, Ziboh A. 2001. Gamma-linolenic acid: Recent advances in biotechnology and clinical applications. AOCS Press, Champaign, IL.

Hussein G, Miyashiro H, Nakamura N, Hattori M, Kakiuchi N, et al. 2000. Inhibitory effects of Sudanese medicinal plant extracts on hepatitis C virus (HCV) protease. Phytotherapy Research 14(7): 510-516.

Joshi K, Lad S, Kale M, Patwardhan B, Mahadik SP, et al. 2006. Supplementation with flax oil and vitamin C improves the outcome of Attention Deficit Hyperactivity Disorder (ADHD). Prostaglandins, Leukotrienes and Essential Fatty Acids 74(1): 17-21.

Kumar PD. 1997. The role of coconut and coconut oil in coronary heart disease in Kerala, South India. Tropical Doctor 27(4): 215-217.

Kuno N, Tsuchiya K, Nakajima S. 1999. Anti-aging cosmetics containing lignan glycosides from germinated sesame seeds. Nippon Keshohin Gijutsusha Kaishi 33(3): 245-253.

Kurita H, Shimogaki H. 2001. Hair preparations containing melanocyte growth promoters and gray hair prevention using the promoters. Japanese Kokai Tokkyo Koho. p. 17.

Kuriyama K, Yoshino S, Hotta H, Takisada M, Senoo M, et al. 1998. Antiaging cosmetics containing lignan glycosides and cell activators and/or moisturizers. Japanese Kokai Tokkyo Koho. p. 15.

Lay CL, Dybing DD. 1989. Linseed. In: Robbelen G, Downey RK, Ashri A, editors, Oil crops of the world: Their breeding and utilization. McGraw Hill, New York. p. 416-430.

Manoj G, Thampi BS, Leelamma S, Menon PV. 2001. Effect of dietary fiber on the activity of intestinal and fecal beta-glucuronidase activity during 1,2-dimethylhydrazine induced colon carcinogenesis. Plant Foods for Human Nutrition 56(1): 13-21.

Miyake T. 1999. Preparation of health foods with black soybeans. Japanese Kokai Tokkyo Koho. p. 5.

Nalini N, Sabitha K, Chitra S, Viswanathan P, Menon VP. 1997. Histopathological and lipid changes in experimental colon cancer: Effect of coconut kernel (*Cocos nucifera* Linn.) and (*Capsicum annum* Linn.) red chili powder. Indian Journal of Medical Research 35(9): 964-971.

Nayar NM. 1984. Sesame. In: Simmonds NW, editor, Evolution of crop plants. Longman, London. p. 231-233.

Nielsen NS, Pedersen A, Sandstrom B, Marckmann P, Hoy C. 2002. Different effects of diets rich in olive oil, rapeseed oil and sunflower-seed oil on postprandial lipid and lipoprotein concentrations and on lipoprotein oxidation susceptibility. British Journal of Nutrition 87(5): 489-499.

Osawa T. 1999. Protective role of dietary polyphenols in oxidative stress. Mechanisms of Ageing and Development 111(2-3): 133-139.

Pastorello EA, Varin E, Farioli L, Pravettoni V, Ortolani C, et al. 2001. The major allergen of sesame seeds (*Sesamum indicum*) is a 2S albumin. Journal of Chromatography 756(1-2): 85-93.

Potterat O. 1997. Antioxidants and free radical scavengers of natural origin. Current Organic Chemistry 1(4): 415-440.

Sankar D, Sambandam G, Ramakrishna Rao M, Pugalendi KV. 2005. Modulation of blood pressure, lipid profiles and redox status in hypertensive patients taking different edible oils. Clinica Chimica Acta 355: 97-104.

Simon P, Chevalier V. 2000. Antiwrinkle cosmetic containing a combination of tensor polymers from synthetic and/or natural origin and dendritic polyesters. European Patent Application. p. 12.

Simopoulos AP. 2002. The importance of the ratio of omega-6/omega-3 essential fatty acids. Biomedicine and Pharmacotherapy 56: 365-379.

Sindurani JA, Rajamohan T. 2000. Effects of different levels of coconut fiber on blood glucose, serum insulin and minerals in rats. Indian Journal of Physiology and Pharmacology 44(1): 97-100.

Sirato-Yasumoto S, Katsuta M, Okuyama Y, Takahashi Y, Ide T. 2001. Effect of sesame seeds rich in sesamin and sesamolin on fatty acid oxidation in rat liver. Journal of Agricultural and Food Chemistry 49(5): 2647-2651.

Sonntag NOV. 1981. Composition and characteristics of individual fats and oils. In: Swern D, editor, Baileys industrial oil and fat products. 4th ed., Vol. 1. Wiley, New York. p. 289-477.

Weiss EA. 1983. Oilseed crops. Longman, New York. p. 282-340.

Yamaha K, Mononobe A, Katata T. 2000. Skin external use agent. Japanese Kokai Tokkyo Koho. p. 7.

Yamashita K. 1997. Synergistic effect of sesame lignans and tocopherols. Food and Free Radicals (Proceedings of the First Symposium 1994): 101-112.

Yun TK. 1999. Update from Asia. Asian studies on cancer chemoprevention. Annals of the New York Academy of Sciences 889: 157-192.

Chapter 5

Beans and Nuts

In the most recent version of the U.S. food pyramid, beans and nuts, which are plant products, are grouped with animal products such as meat and fish. This echoes an earlier classification by the United Kingdom's healthy plate dietary guidelines for planning a meal, for losing weight, or maintaining weight, and gives recommended proportions of each group (Diabetes UK 2007), which put beans and other pulses in the protein group together with meat and fish. This sends a clear message that proteins from plant sources can be as good as proteins from animal sources. It also reflects the stand taken by the American Dietetic Association about vegetarianism that "appropriately planned vegetarian diets are healthful, nutritionally adequate and provide health benefits in the prevention and treatment of some diseases" (Messina and Burke 1997).

BEANS

The term *beans* as used in this category refers to the grain legumes, rather than the vegetable form of the specific plant species. Proteins in legume seeds represent about 20% the dry weight in peas and beans and up to 38% to 40% in soybeans and lupines (Derbyshire et al. 1976; Guéguen and Cerletti 1994). Seeds of legumes are rich in high-quality protein, providing humans with a highly nutritious food resource. Beans are a valuable and sometimes the only source of food protein for people in many parts of the world (Duranti and Guis 1997) and for people who have adopted a vegetarian diet because of political reasons; religious beliefs; esthetic, moral, environmental, or eco-

Asian Crops and Human Dietetics

doi:10.1300/5504_05

nomic concerns; or a desire to consume a more healthy diet. This last is a major factor influencing today's vegetarianism. The beneficial effects of beans in a diet are being reexamined because of the health claims made by proponents. Reports in the scientific literature of bean consumption preventing and treating various diseases are mostly those diseases typical of affluent countries. Many kinds of beans such as the soybean are being studied for protein composition and are being exploited in the modern-day nutraceutical industry. Food proteins not only are a source of amino acids but also may play a bioactive role by themselves or can be the precursors of biologically active peptides with various physiological functions, such as the casein-derived peptides that have demonstrated immunomodulating, antihypertensive, antithrombotic, and opioid activities (Kostyra 1996). Once known as the poor man's meat, beans are now presented as a staple food for vegetarians, and most health organizations encourage their frequent consumption. Therefore, the bean's dietary importance is expected to grow to meet the demand by developing countries for protein, to provide the bean's unique nutrient components, and, in developed countries, to reduce the risks related to consumption of animal food (Duranti and Scarafoni 1999). The U.S. Food and Drug Administration (FDA) approved a proposal to allow claims on food labels that foods containing 6.25 g or more soybean protein per serving reduce the risk of heart disease (USFDA 1999). In particular, a daily soybean protein intake of 25 g was considered beneficial, based on a number of previous clinical observations. Legumes are reported to prevent insulin resistance, digestive tract diseases, and obesity (Duranti 2006).

Leterme (2002) reviewed recommendations from health organizations and states that "grain legumes effectively contribute to a balanced diet and can prevent widely diffused diseases, including type II diabetes and cardiovascular diseases." The hypoglycemic properties of beans are due to its low glycemic index (GI). For example, the GI of kidney beans is 27, lentils 28, chickpeas 33, rice 50, whole-meal bread 77, and baked potatoes 85 (Foster-Powell and Miller 1995). The GI is a carbohydrate's potential for raising blood glucose compared with a reference food such as white bread or glucose, whose GI is 100. An international table of GI values has been developed for reference (Foster-Powell et al. 2002).

Many of the beans belong to the Fabaceae, or Leguminosae, the most important of the Dicotyledonae families (Harborne 1994). This family consists of about 18,000 species classified into approximately 650 genera (Polhill and Raven 1981). Since beans belong to Leguminosae, they are also commonly referred to as legumes, or pulses. Legumes are second only to the cereals as food crops cultivated in world agriculture. Although legumes such as beans, soybean, lentils, peas, and chickpeas are widely known and consumed as popular foods, others are consumed only locally in the places where they are grown. Legumes are important food crops not only because of their high-protein and essential amino acid content but also because of their ability to fix nitrogen in soil, which allows them to grow well in and adapt to poor soil conditions. It has long been recognized that legumes are functional foods that can promote good health and have specific therapeutic properties (Geil and Anderson 1994). Legumes have been associated with long-lived food cultures such as the Japanese (soy, tofu, natto, and miso), the Swedish (brown beans, and peas), and the Mediterranean (lentils, chickpeas, and white beans). In a longitudinal study Blackberry et al. (2004) found a higher legume intake was the "most protective dietary predictor of survival amongst the elderly, regardless of their ethnicity."

However, legumes contain antinutritional factors, such as proteinase inhibitors, lectin, raffinose oligosaccharides, saponins, polyphenols, and phytate. Antinutritional factors are plant-defense compounds that are really a result of evolutionary adaptations of plants. In the human diet, antinutritional factors lower the nutritional value of a food by lowering the digestibility, or bioavailability, of nutrients. For example, zinc deficiency prevents normal childhood growth and greatly weakens the immune system, leading to more infections. Iron deficiency retards normal brain development in infants, increases premature deliveries, and increases morbidity of mother and child at or around the time of childbirth. Despite iron, zinc, and other mineral content being generally high in legumes, antinutritional factors decrease their bioavailability and nutritional iron. Nutritional iron and zinc deficiency is most prevalent and severe in populations subsisting predominantly on cereal and legume diets (International Nutritional Anemia Consultative Group 1982). But some defense compounds have been identified in a number of plant species as promoting health and preventing diseases (Duranti et al. 2004; Scarafoni et al. 2004).

This chapter brings attention to some legumes of Asian origin that are in common use in Asia and Africa but less well known in Europe and North America. These are potential candidates for healthy foods of the future.

*Red Gram (*Cajanus cajan, C. indicus; *Fabaceae)*

Red gram is a major grain legume of the tropical world. It is the fifth-most important pulse in the world (Whiteman et al. 1985), with a global production in 2005 of 3,277,995 Mt (FAOSTAT 2006). Also called pigeon pea, it is popular in India as *tuvar dal,* or simply *dal.* It is cultivated throughout India and is an important part of the daily diet. Although the split seeds are the most commonly consumed part of this plant, the stem serves as an important feed for animals and is sometimes used as human food in Africa, India, and the Caribbean. Gram is cooked with vegetables, green leaves, yam, maize, dried cocoyam (taro) grits, and sweet potatoes in oil with spices.

It is used to treat diabetes in India (Grover et al. 2002; Milliken 1997). The red gram is used as an antipyretic, to stabilize menstrual periods, and to treat dysentery in South America (Duke and Vasquez 1994), and it is used for hepatitis and measles in Africa (Abbiw 1990). It contains about 20.4% protein (Thekoronye and Ngoddy 1985).

Gram is an erect annual reaching 2 to 5 m (6 to 16 ft), with a deep and quick-growing taproot. The stems are pubescent and angular; the leaves are trifoliate, alternate, and set in a spiral along the stem. Leaflets are oblong, lanceolate, and pubescent. The flowers are yellow and sometimes streaked purple or red. The pods are flat with an acuminate tip, pubescent, 5 to 10 cm long, 10 to 12 mm wide, and contain two to nine seeds in shades of brown, red, or black.

A hypolipidemic effect was reported in earlier studies (Prema and Kurup 1973). Diets rich in red gram have shown significant hypoglycemic effects in healthy human volunteers (Panlasigui et al. 1995).

*Chickpea (*Cicer arietinum; *Fabaceae)*

Also called Bengal gram or Egyptian pea, chickpea is native to western Asia. In 2005 global chickpea production was 9,171,930 Mt (FAOSTAT 2006). Chickpea is an annual, deep-rooted plant 2 to 2.5 m (6-8 ft) tall, bearing erect or spreading stems with compound leaves.

Chickpeas are called *channa* in India and can be of two types: pale large seeds called *kabuli channa* or darker smaller seeds called *black channa* or *desi channa*. *Desi* types are traditionally grown in India and many parts of Asia and Africa, and they account for more than 80% of the world production of chickpea (Muehlbauer et al. 1982).

The seeds of the plant and sometimes the young shoots and pods are used for human consumption and the rest of the plant as fodder. Chickpea seed contains 13% to 33% protein, 40% to 55% carbohydrate, and 4% to 10% oil (Stallknecht et al. 1995). Fatty acid composition varies with chickpea type but is approximately 50% oleic and 40% linoleic (Duke 1981). Chickpea seeds are consumed fresh as a green vegetable or parched, fried, roasted, and boiled; as a snack, sweet, or condiment. They are also ground into flour and used to make soup, bread, and sweetmeats. Chickpeas are commonly found in salad bars in the United States and are extensively used for making dishes such as couscous, hummus, and falafel in the Mediterranean region, the Middle East, and parts of Asia.

*Fava Bean (*Vicia faba; *Fabaceae)*

Also called broad bean, horse bean, and faba bean, the fava bean is native to central Asia and the eastern Mediterranean. It has long been cultivated in Asia and Africa and has spread widely to many parts of the world. The plants are bushy and grow 1 to 1.2 m (3-4 ft) tall, bearing dull white flowers and light green pods. The pods are 15 to 30 cm (6-12 in) long and contain five to seven large flat seeds. The nutrient composition of fava beans is 26% protein, 1% to 2% fat, 58% carbohydrates, and 25% dietary fiber (U.S. Department of Agriculture 2001). Fava beans have a relatively high GI compared with other legumes, being 79 ± 16% of the glucose standard (Foster-Powell and Miller 1995). Fava beans have shown a beneficial effect on lipid profiles (Macarulla et al. 2001; Weck et al. 1983), and a cholesterol-reducing efficacy comparable with that of soy protein. Water-soluble proteins isolated from fava beans act as free radical scavengers and enhance the antioxidant capacity in model tissues (Okada and Okada 2000).

*Lentil (*Lens esculenta; *Fabaceae)*

The lentil originated in central Asia and the Middle East. It is a bushy annual grown for its lens-shaped seeds, whose color varies from yellow, orange, red, green, or brown to black. Lentils are preferentially used in place of red gram because they have a shorter cooking time. The plants grow 50 to 140 cm (2-5 ft) tall and the seeds grow in pods, usually two seeds in each. The seeds contain about 25% protein and are an important protein source for people living in Africa and India. A comparative nutritional value of lentil seeds and selected legumes is summarized in Table 5.1 (Iqbal et al. 2006). Lukito (2001) called the lentil a "candidate food for cardiovascular protection," Fernandez-Orozco et al. (2003) described the antioxidant properties of lentils, and Jain et al. (1999) documented the antioxidant properties of lentils and an inverse relationship of lentil consumption to prostate cancer risk.

*Adzuki Bean (*Phaseolus angularis; *Fabaceae)*

The adzuki bean originated in China. It is popular in Japan, China, and Korea, where it is consumed as a paste, *an* or *ann,* or as boiled and sweetened whole beans and used in desserts, snacks, and confectionery. The plant is erect, 0.2 to 0.6 m (1-2 ft) tall, with compound leaves. The plants bear yellow flowers and a cluster of several smooth, short, small, cylindrical pods. The seeds are small and red and are

TABLE 5.1. Mineral Constituents of Important Grain Legumes

	Chickpea	Cowpea	Lentil	Greenpea
Crude protein (g/100 g)	24.0	24.7	26.1	24.9
Minerals (mg/100 g)				
Sodium	101	102	79	111
Potassium	1,155	1,280	874	1,021
Phosphorus	251	303	294	283
Calcium	197	176	120	110
Iron	3.0	2.6	3.1	2.3
Copper	11.6	9.7	9.9	10.0
Zinc	6.8	5.1	4.4	3.2
Manganese	1.9	1.7	1.6	2.2
Magnesium	4.6	4.8	4.5	4.2

Source: Iqbal et al. 2006.

used primarily as a dry bean for sprouts, whole bean, or bean meal, but many cooks use them green. The tender pods are also cooked as a fresh vegetable. Since the seeds have a sweeter taste than most beans, they are often used in desserts and as sweet fillings in confectioneries. Adzuki beans are described as antihyperglycemic (Itoh, Kita et al. 2004) and as having cancer-preventing properties (Itoh, Itoh et al. 2004).

*Cowpea (*Vigna unguiculata, V. sinensis; *Fabaceae)*

The cowpea originated in Asia and is one of the most ancient crops known to humans. It is cultivated around the world primarily as a pulse whose peas are used for human food and animal feed and for its green leaves which are used as a vegetable. It is grown widely in Africa (Niger, Nigeria, Tanzania, Mali, and Senegal), Brazil, the West Indies, Europe, Australia, the United States, and in Asia, especially in India, Myanmar, and Sri Lanka. Annual dry seed production is estimated at 1.24 million tons. In 2005 global dried cowpea production was 3,697,402 Mt (FAOSTAT 2006). The cowpea is a warm-season crop that can tolerate heat and drought. Grown as a vine or bush, the bush type grows 0.6 to 1 m (2-3 ft) tall with leaves that are alternate and trifoliate. The flowers are arranged in racemose or intermediate inflorescences at the distal ends of peduncles that are 5 to 60 cm long. They are white, dirty yellow, pink, pale blue, or purple. The pods are 7.5 to 17.5 cm (3-7 in) long and have 6 to 13 seeds. Young leaves, green pods, and green seeds are used as vegetables; dry seeds are used in a variety of food preparations. Cowpea seeds are also processed and sold dried, frozen, and canned. The pods are processed into paste or flour and used as a food ingredient or starting material for a variety of local foods in Africa. The cowpea is widely recognized as a potential source of protein and other nutrients. It is cultivated for its immature pods and mature seeds and is consumed by people all around the world, especially in developing nations (Phillips and McWatters 1991; Uzogara and Ofuya 1992).

Consumption of processed cowpea seeds with cooked rice, sorghum, or pearl millet is a common practice among the rural people in India. Light decoctions of the seeds are consumed as a remedy for the common cold. The whole seeds have been reported to contain 0.18% to 0.59% tannins (Reddy et al. 1985), phenolic acids (such as *p*-hydroxybenzoic acid, protocatechuic acid, and 2,4-dimethoxybenzoic acid), and cinnamic acid derivatives (such as *p*-coumaric acid), caffeic acid, and ferulic acid (Cai et al. 2003; Sosulski and Dabrowski 1984). Other studies have established the potent antioxidant activities of the raw and processed seeds (Siddhuraju and Becker 2007).

*Mung Bean (***Phaseolus aureus;** *Fabaceae)*

The mung bean, or green gram, is an ancient crop of Asia, having many seed sizes, shapes, and colors. It is an erect bushy annual widely cultivated in warm regions of India, Indonesia, and the United States for forage and especially its edible seeds. The plants are an erect, bushy annual growing 0.3 to 1 m (1 to 3 ft) tall, producing pale yellow flowers and clusters of slender fuzzy pods 5 to 6 in long with small brown or green seeds. The seeds are sprouted and consumed by the Chinese, the Japanese, and Indians as a regular dietary component. Mung beans are similar in composition to other members of the legume family, with 24% protein, 1% fat, 63% carbohydrate, and 16% dietary fiber (U.S. Department of Agriculture 2001). Similar to most legumes, mung beans lack sulfur-containing amino acids, and uncooked beans contain trypsin inhibitors (Khader and Rao 1996). Cooking slightly improves the nutritive value. Mung bean starch is considered to be a low-GI carbohydrate (Juliano et al. 1989; Kabir et al. 1998, 2000; Lerer-Metzger et al. 1996). Dietary fiber prepared from crude mung bean sprouts (Nishimura et al. 2000) produced a significant reduction in total plasma cholesterol levels. Mung beans also showed antioxidant (Duh et al. 1999; Lee et al. 2000a,b) and hepatoprotective (Wu and Ding 2001) properties.

*Black Gram (***Phaseolus mungo;** *Fabaceae)*

The black gram originated in India where it has been in cultivation from ancient times, becoming one of the most highly prized pulses of India. It is an erect and spreading, or suberect and trailing, densely hairy annual herb. The pods are narrow, cylindrical, and up to 6 cm long. It is the most nutritious of all pulses and is recommended for diabetics. It is boiled and eaten whole or after splitting into dal. It is extensively used in various culinary preparations like *idli,* curries, and *papad.* Though very beneficial in limited quantities, excessive intake causes flatulence, which can be prevented by adding a little asafetida, pepper, and ginger. This bean plant is similar to mung bean except that plants are more prostrate, pods are long and hairy, and seeds are oblong and black. Studies have shown that black gram fiber exhibits significant hypoglycemic action in experimental animals (Boby and Leelamma 2003).

Soybean (Glycine max; *Fabaceae*)

The soybean is of east Asian origin and cultivated around the world. In 2005 global soybean production was 209,975,643 Mt (FAOSTAT 2006). Plants are a bushy annual, growing to 2 m (6.6 ft) tall. The leaves are pinnately trifoliolate; flowers are white, pink, greenish blue, violet, or purple; pods are oblong, pendant, 2 to 8 cm long, 0.5 to 2 cm wide, with two to four yellowish brown seeds per pod. Soybeans are not always classified as a legume because they are cultivated more as a source of edible oil. However, because they are an excellent source of protein they are included in this chapter. The soybean is well known in all parts of the world and thus cannot be considered an underutilized legume like the others. A number of publications exist that are dedicated exclusively to soybeans. Because of its exceptional nutritional qualities and the ease of crop production and processing, it has been studied widely. Dry soybean contains 35% protein; 19% oil; 28% carbohydrate; and 5% minerals, B vitamins, and tocopherols. Soybeans are processed into several products such as flour, tofu, milk, tempe, miso, sauce, and protein isolates. Detailed nutritional values of these prepared soy products are summarized by Sugano (2005). Major health benefits attributed to soy consumption include prevention of osteoporosis and cardiovascular diseases, regulation of lipid metabolism, and promotion of blood coagulation. Soy has shown hypocholesterolemic, antiatherogenic, anticarcinogenic, and antiallergenic properties (Sugano 2005). Sugano described soybean as a "treasure box of functionality" because of all the health benefits attributed to soybean and its products.

Velvet Bean (Stizolobium hassjoo; *Fabaceae*)

Velvet beans originated in India and were probably introduced into the Western Hemisphere via Mauritius. Velvet beans are grown in the tropics and subtropics (the Philippines, India, Nigeria, Ghana, Brazil,

Sudan, and Malawi). In folk medicine their uses include as an aphrodisiac (Jayaweera 1981), emetic (Duke 1981), treatment for nervous disorders (Jayaweera 1981; Wijeratne 1987), and arthritis remedy (Wijeratne 1987). The bean paste is also reported to have been used to treat scorpion stings (Jayaweera 1981). Aqueous extracts of leaves show antivenin activity (Houghton and Skari 1994). It is an herbaceous annual that can be grown as a bush or as a perennial climbing vine that can reach 20 m long. Stems are slender and pubescent, bearing alternate trifoliate glabrous leaves. The flowers are purple, red, or greenish yellow. The fruit pods are 10 to 20 cm long, hard, curved, slightly ridged, and covered with soft black, white, or gray hairs, which give the pods a velvety appearance. The pods contain four to six globular seeds that can be brown, gray, white, or black.

Although it is a good source of protein and unsaturated fatty acids, velvet bean also contains a number of toxic elements such as tannins, L-dopa, lectins, and protease inhibitors. However, it is used as human food in Africa after removal of the seed coats and careful preparation to remove these toxins. The pods are also fed to livestock. The seeds are boiled, roasted, or fermented to produce a curd, or bean cake. The beans are also processed into flour and used in making bread. The mature seeds contain 314.4 g/kg crude protein, 51.6 g/kg crude fiber, 67.3 g/kg crude fat, 41.1 g/kg ash, and 525.6 g/kg carbohydrate as reported by Siddhuraju et al. (1996), who did a complete biochemical analysis of this underused legume. In a similar study, Vadivel and Janardhanan (2000) indicated the increased attention this legume has received in the past few years. The possibility of obtaining L-dopa, which is relatively expensive to synthesize and is used to treat Parkinson's disease, from velvet beans has received attention in recent years (Pugalenthi et al. 2005).

*Rice Bean (*Phaseolus calcaratus, V. umbellate, V. calcarata; *Fabaceae)*

The rice bean is native to Southeast Asia and is grown in China, Fiji, India, Malaysia, Nepal, and the Philippines. The plant is a perennial but grown as an annual vine or bush bearing yellow flowers and small, slender, edible beans. It grows 0.6 to 1 m (2-3 ft) tall and bears bright yellow flowers followed by the appearance of clusters of pods in the leaf axils. Pods are 10 to 12 cm long, with 7 to 10 smooth seeds per pod. The bean is frequently cooked with curry as a separate dish, with rice, or in stews and soups. The green immature pods and the young leaves are edible but not usually consumed as vegetables. Srivatsava et al. (2001) reported that this underutilized legume was comparable in quality to other traditional pulses containing 3.2% to 4.43% crude fiber. They studied the nutritional quality of 17 genotypes and found rice beans to contain (per 100 g) 290 to 440 mg phosphorus, 2.69 to 3.37 mg zinc, 0.55 to 1.2 mg copper, and 5.46 to 7.49 mg iron.

NUTS

Nuts are good sources of proteins and antioxidant compounds. The documented inverse association of nut consumption and risk of coronary heart disease by lipid modulation has garnered interest from the scientific and lay community.

*Almond (***Amygdalus communis, Prunus dulcis;** ***Rosaceae)***

The almond is native to Asia and Africa and is cultivated around the world for its nutritious nuts which are rich in protein. In 2005 global almond production was 1,619,906 Mt (FAOSTAT 2006). Almond trees grow well in tropical and subtropical regions, usually reaching 6 to 9 m (20-30 ft), with spreading branches and lanceolate leaves finely serrated at the edges. The flowers are pink or white. The almond nut, with or without the brown skin, is consumed whole or used in various confectioneries and chocolates. Almond oil is a rich source of antioxidants, with alpha-tocopherol at 24.2 mg/100 g of extracted oil. Apart from its nutritional value, the almond is reported to have beneficial effects on levels of blood cholesterol and lipoprotein (Spiller et al. 1998). Diets containing almond meal or oil have been shown to reduce colon cancer risk in rats (Davis and Iwahashi 2001) and significantly reduce plasma triacylglycerols and total and low-density-lipoprotein cholesterol while increasing high-density-lipoprotein levels in humans (Hyson et al. 2002). In addition, almonds when used as snacks and in the diet of hyperlipidemic subjects, significantly reduced coronary heart disease factors (Kendall et al. 2002). Long-term almond supplementation showed spontaneous nutrient modification of an individual's habitual diet that closely matched the recommendations to prevent cardiovascular and other chronic diseases (Jaceldo-Siege et al. 2004). Extracts of whole almond seed, the brown skin, and green shell cover possess potent free-radical-scavenging capacities (Siriwardhana and Shahidi 2002). These activities may be related to the presence of flavonoids and other phenolic compounds in nuts. Almond hulls have been shown to serve as a rich source of triterpenoids, betulinic, urosolic, and oleanolic acids (Takeoka et al. 2000), as well as flavonol glycosides and phenolic acids (Sang et al. 2002b). Sang et al. (2002a) isolated, identified, and characterized other components of the almond nut to precisely estimate its nutraceutical values. They found catechin, protocatechuic acid, vanillic acid, *p*-hydroxybenzoic acid, naringenin glucoside, galactoside, glucoside, rhamnoglucoside of 3′-*O*-methylquercetin, and rhamnoglucoside of kaempferol.

Pistachio (**Pistacia vera**; *Anacardiaceae*)

Pistachio is a popular flavoring. Nut consumption is inversely associated with risk markers of coronary heart disease (lipids). In 2005 global pistachio production was 501,237 Mt and the area harvested was 448,238 ha (FAOSTAT 2006). It is a small tree up to 12 m tall, and native to Asia. The leaves are pinnate, 10 to 20 cm long, and the flowers are apetalous, unisexual, and borne in panicles. The fruit is a drupe, containing an elongated seed, which is the green pistachio nut. When the fruit ripens, the shells partially split open and the kernels are eaten whole, fresh or roasted, and salted. They are used to flavor confectioneries and a number of prepared foods, desserts, and rice dishes. The pistachio is a good source of protein (22%), fat (55%), and carbohydrate (19%). The mean mineral content of pistachio nuts is (mg/100 g) sodium, 9.38; magnesium, 116.90; potassium, 642.07; calcium, 171.69; copper, 0.75; zinc, 2.77; and iron, 0.58 (Cağlarirmak and Batkan 2005). Pistachio oil contains the antioxidant lutein at 4.4 mg/100 g of expressed oil.

Walnut (**Juglans sinensis**; *Juglandaceae*)

The walnut is a deciduous tree of Asian origin, growing 12 to 40 m (39-131 ft) tall, with pinnate leaves, 25 to 100 cm long, and bearing monoecious male flowers (catkins) and purple female flowers. In

2005 global walnut production was 1,507,567 Mt and the area harvested was 629,748 ha (FAOSTAT 2006).

Walnuts contribute nutrients essential to health. Eating walnuts is one of the easiest things a person can do to improve his or her health. Many relish their taste, making walnuts ideally suited for inclusion in any diet as part of meals or snacks. Walnuts can be eaten with fresh or dried fruit; in salads, soups, yogurt, breakfast cereals, or oatmeal; or with rice and pasta dishes. Dietetic professionals and diabetes educators encourage type 2 diabetes patients to incorporate walnuts in a low- to moderate-fat diet. "Walnuts are unique because they have a perfect balance of n-6 and n-3 polyunsaturated fatty acids, a ratio of 4:1, which [the Lyon Heart Study has] . . . shown to decrease the risk of sudden death" (Simopoulos 2002).

The heart's benefits of walnuts include lowering cholesterol, increasing the ratio of high-density-lipoprotein cholesterol to total cholesterol, reducing inflammation, and improving arterial function (Ros et al. 2004; Tapsell et al. 2004; Zhao et al. 2004). A number of other studies also have documented the lipid-lowering effect of walnut in the diet (Iwamoto et al. 2002; Mukuddem-Petersen et al. 2005; Ros et al. 2004; Sabate et al. 1993; Zambon et al. 2000).

REFERENCES

Abbiw DK. 1990. Useful plants of Ghana. Royal Botanic Gardens, Richmond, Surrey, UK.

Blackberry I, Kouris-Blazos A, Wahlqvist ML, Steen B, Lukito W, et al. 2004. Legumes: The most important dietary predictor of survival in older people of different ethnicities. Asia Pacific Journal of Clinical Nutrition 13(Suppl.): S126.

Boby RG, Leelamma S. 2003. Blackgram fiber (*Phaseolus mungo*): Mechanism of hypoglycemic action. Plant Foods for Human Nutrition 58(1): 7-13.

Cağlarirmak N, Batkan AC. 2005. Nutrients and biochemistry of nuts in different consumption types in Turkey. Journal of Food Processing and Preservation 29(5-6): 407-423.

Cai R, Hettiarachchy NS, Jalaluddin M. 2003. High-performance liquid chromatography determination of phenolic constituents in 17 varieties of cowpeas. Journal of Agricultural and Food Chemistry 51: 1623-1627.

Davis PA, Iwahashi CK. 2001. Whole almonds and almond fractions reduce aberrant crypt foci in a rat model of colon carcinogenesis. Cancer Letters 165: 27-33.

Derbyshire E, Wright DJ, Boulter D. 1976. Legumin and vicilin, storage proteins of legume seeds. Phytochemistry 15: 3-24.

Diabetes UK. 2007. Planning your meals. http://www.diabetes.org.uk/Guide-to-diabetes/Food_and_recipes/Food_and_diabetes/Planning_your_meals/. Accessed June 1, 2007.

Duh PD, Du PC, Yen GC. 1999. Action of methanolic extract of mung bean hulls as inhibitors of lipid peroxidation and non-lipid oxidative damage. Food and Chemical Toxicology 37: 1055-1061.

Duke JA. 1981. Handbook of legumes of world economic importance. Plenum Press, New York.

Duke JA, Vasquez R. 1994. Amazonian ethnobotanical dictionary. CRC Press, Boca Raton, FL.

Duranti M. 2006. Grain legume proteins and nutraceutical properties. Fitoterapia 77(2): 67-82.

Duranti M, Gius C. 1997. Legume seeds: Protein content and nutritional value. Field Crops Research 53: 31-45.

Duranti M, Scarafoni A. 1999. Modification of storage protein content and quality in legume seeds. Journal of New Seeds 1(1): 17-35.

Duranti M, Lovati MR, Dani V, Barbiroli A, Scarafoni A, et al. 2004. The alpha' subunit from soybean 7S globulin lowers plasma lipids and upregulates liver beta-VLDL receptors in rats fed a hypercholesterolemic diet. Journal of Nutrition 134: 1334-1339.

FAOSTAT (Statistics Division of Food and Agriculture Organization of the United Nations). 2006. http://faostat.fao.org. Accessed February 2006.

Fernandez-Orozco R, Zielinski H, Piskula MK. 2003. Contribution of low-molecular-weight antioxidants to the antioxidant capacity of raw and processed lentil seeds. Nahrung 47(5): 291-299.

Foster-Powell K, Holt SHA, Miller JB. 2002. International table of glycemic index and glycemic load values: 2002. American Journal of Clinical Nutrition 76(1): 5-56.

Foster-Powell K, Miller JB. 1995. International tables of glycemic index. American Journal of Clinical Nutrition 62: 871S-890S.

Geil PB, Anderson JW. 1994. Nutrition and health implications of dry beans: A review. Journal of the American College of Nutrition 13: 549-558.

Grover JK, Yadav S, Vats V. 2002. Medicinal plants of India with anti-diabetic potential. Journal of Ethnopharmacology 81: 81-100.

Guéguen J, Cerletti P. 1994. Proteins of some legume seeds, soybean, pea, fababean and lupin. In: Hudson BJF, editor, New and developing sources of food proteins. Chapman and Hall, New York. p. 145.

Harborne JB. 1994. Phytochemistry of the leguminosae. In: Bisby FA, Southon IW, editors, Phytochemical dictionary of the leguminosae. Chapman and Hall, Boca Raton, FL. p. 20-24.

Houghton PJ, Skari KP. 1994. The effect on blood clotting of some West African plants used against snakebite. Journal of Ethnopharmacology 44: 99-108.

Hyson D, Schneeman BO, Davis PA. 2002. Almonds and almond oil have similar effects on plasma lipids and LDL oxidation in healthy men and women. British Journal of Nutrition 132: 703-707.

International Nutritional Anemia Consultative Group. 1982. Iron Absorption from Cereals and Legumes. A Report of the International Nutritional Anemia Consultative Group. Nutrition Foundation, New York. p. 1-44.

Iqbal A, Khalil IA, Ateeq N, Khan MS. 2006. Nutritional quality of important food legumes. Food Chemistry 97(2): 331-335.

Itoh T, Itoh Y, Mizutani M, Fujishiro K, Furuichi Y, et al. 2004. Hot-water extracts from adzuki beans (*Vigna angularis*) suppress not only the proliferation of KATO III cells in culture but also benzo(a)pyrene-induced tumorigenesis in mouse forestomach. Journal of Nutritional Science and Vitaminology (Tokyo) 50(4): 295-299.

Itoh T, Kita N, Kurokawa Y, Kobayashi M, Horio F, Furuichi Y. 2004. Suppressive effect of a hot water extract of adzuki beans (*Vigna angularis*) on hyperglycemia after sucrose loading in mice and diabetic rats. Bioscience, Biotechnology, and Biochemistry 68(12): 2421-2426.

Iwamoto M, Imaizumi K, Sato M, Hirooka Y, Sakai K, et al. 2002. Serum lipid profiles in Japanese women and men during consumption of walnuts. European Journal of Clinical Nutrition 56: 629-637.

Jaceldo-Siege K, Sabate J, Rajaram S, Fraser GE. 2004. Long-term almond supplementation without advice on food replacement induces favorable nutrient modifications to the habitual diets of free-living individuals. British Journal of Nutrition 92: 533-540.

Jain MG, Hislop GT, Howe GR, Ghadirian P. 1999. Plant foods, antioxidants, and prostate cancer risk: Findings from case-control studies in Canada. Nutrition and Cancer 34(2): 173-184.

Jayaweera DMA. 1981. Medicinal plants used in Ceylon. National Science Council of Sri Lanka, Colombo, Sri Lanka.

Juliano BO, Perez CM, Komindr S, Banphotkasem S. 1989. Properties of Thai cooked rice and noodles differing in glycemic index in noninsulin-dependent diabetics. Plant Foods for Human Nutrition 39: 369-374.

Kabir M, Guerre-Millo M, Laromiguiere M, Slama G, Rizkalla SW. 2000. Negative regulation of leptin by chronic high-glycemic index starch diet. Metabolism 49: 764-769.

Kabir M, Rizkalla SW, Quignard-Boulange A, Guerre-Millo M, Boillot J, et al. 1998. A high glycemic index starch diet affects lipid storage-related enzymes in normal and to a lesser extent in diabetic rats. Journal of Nutrition 128: 1878-1883.

Kendall CWC, Jenkins DJA, Marchie A, Parker T, Connelly P. 2002. Dose response of almonds on coronary heart disease risk factors-blood lipids, oxidized LDL, Lp(a), homocysteine and pulmonary nitric oxide: A randomized controlled cross-over trial. Circulation 106: 1327-1332.

Khader V, Rao SV. 1996. Studies on protein quality of green gram (*Phaseolus aureus*). Plant Foods for Human Nutrition 49: 127-132.

Kostyra H. 1996. Food proteins: Evolution and nutritional aspects. In: S. Bardocz, E. Gelencsér, A. Pusztai, editors, Effects of antinutrients on the nutritional value of legume diets, COST98, Vol. 1. European Commission Directorate-General XII, Brussels. p. 86.

Lee KG, Mitchell AE, Shibamoto T. 2000a. Antioxidative activities of aroma extracts isolated from natural plants. Biofactors 13: 173-178.

Lee KG, Mitchell AE, Shibamoto T. 2000b. Determination of antioxidant properties of aroma extracts from various beans. Journal of Agricultural and Food Chemistry 48: 4817-4820.

Lerer-Metzger M, Rizkalla SW, Luo J, Champ M, Kabir M, et al. 1996. Effects of long-term low-glycaemic index starchy food on plasma glucose and lipid concentrations and adipose tissue cellularity in normal and diabetic rats. British Journal of Nutrition 75: 723-732.

Leterme P. 2002. Recommendations by health organizations for pulse consumption. British Journal of Nutrition 88(Suppl. 3): 239-242.

Lukito W. 2001. Candidate foods in the Asia-Pacific region for cardiovascular protection: Nuts, soy, lentils and tempe. Asia Pacific Journal of Clinical Nutrition 10(2): 128-133.

Macarulla MT, Medina C, De Diego A, Chàvarri M, Zulet MA, et al. 2001. Effects of the whole seed and a protein isolate of faba bean (*Vicia faba*) on the cholesterol metabolism of hypercholesterolaemic rats. British Journal of Nutrition 85(5): 607-614.

Messina V, Burke K. 1997. Position of the American Dietetic Association: Vegetarian diets. Journal of the American Dietetic Association 97: 1157-1159.

Milliken W. 1997. Plants for malaria, plants for fever. Royal Botanic Gardens, Richmond, Surrey, UK.

Muehlbauer, FJ, Short RW, Kaiser WJ. 1982. Description and culture of garbanzo beans. Coop Ext Publ EB 1112. Washington State University, Pullman.

Mukuddem-Petersen J, Oosthuizen W, Jerling JC. 2005. A systematic review of the effects of nuts on blood lipid profiles in humans. Journal of Nutrition 135(9): 2082-2089.

Nishimura N, Taniguchi Y, Kiriyama S. 2000. Plasma cholesterol-lowering effect on rats of dietary fiber extracted from immature plants. Bioscience, Biotechnology, and Biochemistry 64: 2543-2551.

Okada Y, Okada M. 2000. Effect of a radical scavenger "water soluble protein" from broad beans (*Vicia faba*) on antioxidative enzyme activity in cellular aging. Journal of Nutritional Science and Vitaminology (Tokyo) 46: 1-6.

Panlasigui LN, Panlilio LM, Madrid JC. 1995. Glycemic response in normal subjects to five different legumes commonly used in the Philippines. Journal of Nutritional Science and Vitaminology (Tokyo) 46(2): 155-160.

Phillips R, McWatters K. 1991. Contribution of cowpeas to nutrition and health. Food Technology 45: 127-130.

Polhill RM, Raven PH. 1981. Advances in legume systematics. Royal Botanic Gardens, Kew, UK.

Prema L, Kurup PA. 1973. Hypolipidemic activity of the protein isolated from *Cajanus cajan* in high fat-cholesterol diet fed rats. Indian Journal of Biochemistry and Biophysics 10: 293-296.

Pugalenthi M, Vadivel V, Siddhuraju P. 2005. Alternative food/feed perspectives of an underutilized legume *Mucuna pruriens* var. Utilis: A review. Plant Foods for Human Nutrition 60(4): 201-218.

Reddy NR, Pierson MD, Sathe SK, Salunkhe DK. 1985. Dry bean tannins: A review of nutritional implications. Journal of the American Oil Chemists' Society 62: 541-549.

Ros E, Nunez I, Perez-Heras A, Serra M, Gilabert R, et al. 2004. A walnut diet improves endothelial function in hypercholesterolemic subjects: A randomized crossover trial. Circulation 109: 1609-1614.

Sabate J, Fraser GE, Burke K, Knutsen SF, Bennett H, et al. 1993. Effects of walnuts on serum lipid levels and blood pressure in normal men. New England Journal of Medicine 328: 603-607.

Sang S, Lapsley K, Jeong WS, LaChance PA, Ho CT, et al. 2002a. Antioxidative phenolic compounds isolated from almond skins (*Prunus amygdalus* Batsch). Journal of Agricultural and Food Chemistry 50: 2459-2463.

Sang S, Lapsley K, Rosen RT, Ho CT. 2002b. New prenylated benzoic acid and other constituents from almond hulls (*Prunus amygdalus* Batsch). Journal of Agricultural and Food Chemistry 50: 607-609.

Scarafoni A, Consonni A, Panozzo JF, Magni C, Sironi E, et al. 2004. Proteome analysis of lentil seeds with low trypsin inhibitor content. In: The Proceedings of the 5th Grain Legume Conference, Dijon, France. p. 47-50.

Siddhuraju P, Becker K. 2007. The antioxidant and free radical scavenging activities of processed cowpea (*Vigna unguiculata* (L.) Walp.) seed extracts. Food Chemistry 101(1): 10-19.

Siddhuraju P, Vijayakumari K, Janardhanan K. 1996. Chemical composition and protein quality of the little-known legume, velvet bean (*Mucuna pruriens* (L.) DC.). Journal of Agricultural and Food Chemistry 44(9): 2636-2641.

Simopoulos AP. 2002. The importance of the ratio of omega-6/omega-3 essential fatty acids. Biomedicine and Pharmacotherapy 56: 365-379.

Siriwardhana SSKW, Shahidi F. 2002. Antiradical activity of extracts of almond and its by-products. Journal of the American Oil Chemists' Society 79: 903-908.

Sosulski FW, Dabrowski KJ. 1984. Composition of free and hydrolysable phenolic acids in the flours and hulls of ten legume species. Journal of Agricultural and Food Chemistry 32: 131-133.

Spiller GA, Jenkins AJ, Bosello O, Gates JE, Cragen LN, et al. 1998. Nuts and plasma lipids: An almond-based diet lowers LDL-C while preserving HDL-C. Journal of the American College of Nutrition 17: 285-290.

Srivatsava RP, Srivatsava GK, Gupta RK. 2001. Nutritional quality of rice bean (*Vigna umbellata*). Indian Journal of Agricultural Biochemistry 14(1-2): 55-56.

Stallknecht G, Gilberston KM, Carlson GR, Eckhoff JL, Kushnak GD, et al. 1995. Production of chickpeas in Montana. Montana AgResearch 12: 46-50.

Sugano M. 2005. Soy in health and disease prevention. CRC Press, New York.

Takeoka G, Dao L, Teranishi R, Wong R, Flessa S, et al. 2000. Identification of three triterpenoids in almond hulls. Journal of Agricultural and Food Chemistry 48: 3437-3439.

Tapsell LC, Gillen LJ, Patch CS, Batterham M, Owen A, et al. 2004. Including walnuts in a low-fat/modified-fat diet improves HDL cholesterol-to-total cholesterol ratios in patients with type 2 diabetes. Diabetes Care 27: 2777-2783.

Thekoronye AI, Ngoddy PO. 1985. Integrated food science and technology for tropics. Macmillan, London.

U.S. Department of Agriculture. 2001. Nutrient database for standard reference. Release 14. U.S. Department of Agriculture, Washington, DC.

USFDA (United States Food and Drug Administration). 1999. Final rule. Federal Register 64: 57699.

Uzogara S, Ofuya Z. 1992. Processing and utilisation of cowpeas in developing countries: A review. Journal of Food Processing and Preservation 16: 105-147.

Vadivel V, Janardhanan K. 2000. Nutritional and antinutritional composition of velvet bean, an under-utilized food legume in South India. International Journal of Food Science and Nutrition 51(4): 279-287.

Weck M, Hanefeld M, Leonhardt W, Haller H, Robowsky KD, et al. 1983. Field bean protein diet in hypercholesterolemia. Nahrung 27: 327-333.

Whiteman PC, Byth DE, Wallis ES. 1985. Pigeonpea. In: Summerfield RJ, Roberts EH, editors, Grain legume crops. Collins Professional and Technical Books, London. p. 698-698.

Wijeratne S. 1987. Poisonous plants, another's medicine. Mirror Magazine (Sri Lanka) 18: 8-9.

Wu J, Ding X. 2001. Hypotensive and physiological effect of angiotensin converting enzyme inhibitory peptides derived from soy protein on spontaneously hypertensive rats. Journal of Agricultural and Food Chemistry 49: 501.

Zambon D, Sabate J, Munoz S, Campero B, Casals E, et al. 2000. Substituting walnuts for monounsaturated fat improves the serum lipid profile of hypercholesterolemic men and women. A randomized crossover trial. Annals of Internal Medicine 132: 538-546.

Zhao G, Etherton TD, Martin KR, West SG, Gillies PJ, et al. 2004. Dietary alphalinolenic acid reduces inflammatory and lipid cardiovascular risk factors in hypercholesterolemic men and women. Journal of Nutrition 134: 2991-2997.

Chapter 6

Spices

Although spices are added in such small quantities to flavor foods and drinks that their caloric values are negligible, their important chemical compounds and antioxidant activities can offer health promotion and protection from diseases. Consuming spiced foods often leads to greater thermogenesis and in some cases to greater satiety. Antioxidant spices greatly reduce low-density lipoprotein (LDL) oxidation and modulate the synthesis of prostaglandins and leukotrienes. Several spices or their extracts have also been found to inhibit platelet aggregation. Thus these spices can be considered functional agents and nutraceuticals that could help in preventing diseases and obesity.

ASAFETIDA

Also known as devil's dung and *hing* in Hindi, asafetida (*Ferula asafoetida;* Apiaceae) is a native of Persia, Afghanistan, and several neighboring countries. Asafetida is a spice and a condiment in Asian cooking, especially Indian cuisine, where most dishes are garnished with asafetida fried in hot oil. Indians believe that use of asafetida regularly in the diet enhances the sweetness of the voice and the ability to effortlessly perform vocal music with greater vigor.

Asafetida is a tall perennial herb, grown on a large scale in India, Iran, Afghanistan, and other parts of Asia. It is sometimes called giant fennel, because of its striking similarity to the fennel plant. Asafetida grows 1.8 to 2.4 m (6-8 ft) tall and spreads to 0.6 to 0.9 m (2-3 ft) wide. The stem produces numerous finely divided leaves with petioles covered with wide sheaths. Flowers are rounded umbels on long, stout, branched stalks. The fruit is oval, flat, and thin and has milky

Asian Crops and Human Dietetics

doi:10.1300/5504_06

juice. The roots are fleshy and all parts of the plant have the distinctive fetid smell characteristic of the spice. The edible spice consists of a gum resin made of the milky sap obtained from an incision of the mature root. The plants are harvested just before flowering. The stalks are cut close to the root and the exposed surface is covered with dried leaves or earth. A milky liquid oozes out, which is dried to create the resin that forms the commercial spice. It is available as tears in Chinese pharmacies and in mass, paste, or powder form in other Asian grocery stores. Asafetida is an indispensable condiment for cooking beans and pulses, an essential ingredient in curry, and a desired flavoring for sauces and pickles. It is also used as a flavoring agent, forming an essential constituent of many spice mixtures. The whole plant is sometimes used as a fresh vegetable.

In Ayurvedic medicine it is proposed to have hot energy, which can counteract cold conditions and neurological and digestive disorders. It is a valuable remedy and prescribed for hysteria, nervous disorders, intestinal worms and other parasites, flatulence, emotional stress, nervous palpitations, whooping cough, pneumonia, bronchitis, and asthma. Traditional Chinese medicine suggests that it stimulates the intestinal, respiratory, and nervous system and prescribes it for digestive problems, intestinal parasites, flatulence, and respiratory problems such as asthma, whooping cough, and chronic bronchitis. In some parts of the world, dried asafetida is smoked with powdered adzuki beans to treat asthma. In some parts of Africa and Asia it is administered to pregnant women as a preventive measure against future miscarriages. Other medicinal properties attributed to asafetida include as an anticoagulant, antiseptic, antispasmodic, carminative, expectorant, hypotensive, sedative, and hypolipidemic, and it is used as a nerve stimulant and toothache remedy.

Asafetida is a good source of calcium, iron, dietary fiber, and essential amino acids such as lysine and threonine (Pradeep et al. 1993). Modern research has demonstrated the ability of asafetida to inhibit carcinogenesis, to afford protection against free-radical-mediated diseases (Aruna and Sivaramakrishnan 1990; Saleem et al. 2001; Soudamini et al. 1995), and to inhibit tumors (Unnikrishnan and Kuttan 1990). It is also described as an aphrodisiac (Eigner and Scholz 1990). Other studies show that aqueous extracts of asafetida can inhibit lipid peroxidation (Sujatha and Srinivas 1995) and are hypocholesterolemic, modulating lipid metabolism to decrease cholesterol production (Srinivasan and Srinivasan 1995). Asafetida has also been reported to protect the liver against aflatoxin-induced liver damage by *Aspergillus parasiticus* (Soni et al. 1992) and liver cancer (Soni et al. 1997) and to stimulate digestion (Platel and Srinivasan 1996) by enhancing the activities of digestive enzymes. Furthermore, some studies demonstrate antifungal and antibacterial properties of asafetida and suggest its possible use as a natural food preservative (Thyagaraja and Hosono 1996). Hypotensive antispasmodic effects of asafetida were demonstrated in animal models (Fatehi et al. 2004).

BLACK PEPPER

Black pepper (*Piper nigrum;* Piperaceae) is native to the southwestern coast of India. The pepper plant is widely cultivated throughout tropical Asia, especially Indonesia and India. It is a perennial woody vine with aerial roots that appear at the stem nodes. The leaves are shiny, broad, alternate, and heart shaped. The small white flowers are borne on the vine, at the node opposite the leaves, in slender spikes. Each spike bears about 50 to 100 blossoms. The flowers are hermaphroditic, but some flowers may be unisexual in a spike. The fruits are globular berries, about 5 mm in diameter, green when immature and turning a bright red upon ripening. The fruit is a drupe and bears a single seed. Pepper can be processed to give black, white, green, or red peppercorns, depending on the time of harvest and postprocessing method. The berries are picked when they begin to turn red, immersed in boiling water, and then spread out to dry in the sun for three or four days. White pepper is obtained by removing the outer part of the pericarp before processing. Pepper fruits are harvested when immature and green for use as a vegetable in pickles. Harvesting them at a red-ripe stage obtains red peppercorns. Black pepper is the most commonly used, and is found in nearly every kind of dish all over the world, including some sweet dishes. It is used as a flavoring agent as well as the main spice in many curry dishes and pickles. Whole pepper is used to flavor special dishes and deep-fried items in India.

Pepper has been used in the East for more than 4,000 years for medicinal and culinary purposes. In Chinese medicine white pepper is used to treat malaria, cholera, dysentery, diarrhea, stomachache, and other digestive problems. In many Asian traditional medical practices, pepper is proposed to improve blood circulation and relieve muscular aches and pains, and it is used for rheumatism, sprains, stiffness, neuralgia, and poor muscle tone. Pepper also aids digestion, improves appetite, and is used for colic, diarrhea, flatulence, heartburn, indigestion, loss of appetite, and nausea. It is a popular treatment for colds, coughs, flu, infections, and viruses. Ground pepper taken in combination with turmeric and milk is a home remedy for the common cold and cough. It is used in medicinal formulas as an analgesic, antimicrobial, antiseptic, antispasmodic, antitoxic, purgative, aphrodisiac, bactericidal, carminative, diaphoretic, digestive, diuretic, antipyretic, laxative, and rubefacient. It is sometimes used as a nervous, circulatory, and digestive system stimulant, as well as a stomachic and general tonic.

Pepper contains numerous chemical compounds, including piperine, chavicine, piperidine, and piperettine. Recent studies show that pepper in the diet can offer protection from colon cancer by enhancing beneficial bacterial enzyme activity in the intestine (Nalini et al. 1998). Other studies demonstrated antimicrobial, antibacterial (Dorman and Deans 2000; Pradhan et al. 1999), and antioxidant (Dorman et al. 2000) properties offering protection against mutagenesis (Milic and Milic 1998; Nakatani et al. 1986). Chemical extractions of pepper containing the active ingredients piperine or piperidine are used in numerous commercial medicinal preparations to reduce pain and improve joint function (Hastings et al. 2001) and blood circulation (Chang 1999). Dietary intake of black pepper was found to reduce the risk of atherosclerosis markedly by virtue of its hypolipidemic and antiatherogenic effects in animal models (Vijayakumar et al. 2002). Others reported the antioxidant and antibacterial potential of pepper extracts against *Staphylococcus aureus, Bacillus cereus,* and *B. subtilis* (Singh et al. 2005; Vijayakumar et al. 2004).

In 2005 global pepper production was 409,899 Mt and the area harvested was 471,191 ha (FAOSTAT 2006).

CARDAMOM

Cardamom (*Elettaria cardamomum;* Zingiberaceae) is an aromatic spice valued for its fruits and seeds. Cardamom originated in southern India and Sri Lanka and has been in use as early as the fourth century BC as a medicinal herb. Cardamom was known as *Amomum cardomonum* before its genus name was changed to *Elettaria,* derived from its Hindi name *elaichi.* Cardamom is cultivated in India, Sri Lanka, Jamaica, Nepal, Guatemala, Mexico, Thailand, Central America, and other tropical and subtropical parts of Asia, Africa, and the Pacific islands. Cardamom is a perennial bush with an aerial pseudostem (sheaths of leaves forming a trunklike stem) growing 1.5 to 4.6 m (5-15 ft). The plant has a large rhizome and long, dark green leaves approximately 0.3 to 0.6 m (1-2 ft) long and 10 to 15 cm (4-6 in) wide that grow from the plant base at ground level. The inflorescence is a long panicle with clusters of flowers arising from the underground stem. The flowers are fragrant, small with greenish petals and a purple-veined white tip. The fruit is a capsule with three loci containing 30 to 50 seeds. The fruits are 2 to 4 cm long, ovoid and triangular, and brown or pale pink when ripe.

Medicinally, it has been used frequently as an antinauseant, to freshen breath, and to cure digestive problems. According to Ayurvedic medicine, cardamom reduces the air and water elements and increases appetite. Cardamom is used in medicinal preparations to relieve flatulence and for strengthening digestive function. The Unani system of medicine uses cardamom frequently in the treatment of gastrointestinal disorders and as a stomachic, digestive, and antiemetic. In India, cardamom seeds or powder is used to flavor *pan supari.* Other uses of cardamom include as an aphrodisiac, as a cure for obesity, and to flavor tea and pastries and other dishes.

Cardamom is lavishly used in Indian, Arabic, and Mogul cuisine and its sweet fragrance is frequently used to flavor tea, coffee, and other beverages and desserts. Cardamom is also used in curried dishes and in baked goods. Cardamom is an expensive spice and forms an essential component of *masala* tea, a luxury to most people in India. Inclusion of cardamom in the diets of rats showed enhanced antioxidant enzyme activity and lipid peroxidation activity (Dhuley 1999). Cardamom oil has been reported to be anti-inflammatory and antispasmodic (Al-Zuhair et al. 1996). Other medicinal properties of cardamom demonstrated in animal models include as an antimicrobial (Isao et al. 1991; Ramadan et al. 1994), antifungal (Rahman et al. 2000), antiviral (Ghazi et al. 2000), and antiulcerogenic by protecting the intestinal mucosal membranes (Jafri et al. 2001).

CINNAMON

Cinnamon (*Cinnamomum verum, C. zeylanicum, C. cassica;* Lauraceae) is native to Sri Lanka and India. It is an evergreen tree with a thick, reddish brown bark. The leaves are leathery, ovate to ovate-lanceolate with short petioles, and 10 to 15 cm long. The flowers are panicles with very small yellowish flowers. These panicles bear pointed black fruits from which cinnamon oil is extracted. The fruit is an ellipsoid berry. Cinnamon sticks are made from the bark of the tree.

The use of cinnamon bark dates back thousands of years, to at least 3000 BC. Chinese as well as the Indian Ayurvedic texts from this time mention it as a treatment for fever, diarrhea, and menstrual problems. Cinnamon is cultivated in tropical regions in the Philippines, the West Indies, India, Sri Lanka, China, Madagascar, Brazil, Japan, Taiwan, and the Caribbean. Cinnamon has numerous everyday uses in the kitchen as well as medicinal uses and it is important in cosmetic and perfumery industries. Cinnamon is used in both sweet and savory dishes, rice dishes, desserts, and baked foods, and to flavor beverages, wine, and many processed foods. It is an ingredient in many commercial products, including toothpaste, mouthwash, and other oral hygiene products, because it fights tooth decay and gum disease. It is used as an essential ingredient in many antiseptic formulas that help kill bacteria, fungi, and viruses, and it finds a place in cosmetics, perfumery, and toiletries, including perfume, soap, acne medications, lipstick, chewing gum, and cough syrup. It is popular as a valuable food flavor and spice and is added to carbonated beverages and chocolate drinks.

It is a recommended remedy for nausea, vomiting, diarrhea, indigestion, the common cold, digestive problems, backache, and menstrual problems. The medicinal properties attributed to cinnamon include as an antibacterial, antiseptic, antiviral, antispasmodic, and antifungal. The active compounds in cinnamon are cinnamaldehyde and eugenol. Cinnamaldehyde is a sedative and analgesic. Eugenol has pain-relieving qualities and is used to treat nausea, vomiting, diarrhea, stomach ulcers, acid indigestion, heartburn, lack of appetite, and abdominal disorders. In Chinese medicine, cinnamon bark is believed to have a warming effect on the body and is used for conditions caused by excessive coldness. Cinnamon bark is also prescribed to control blood sugar levels in diabetics, to promote menstruation, and to treat menstrual pain. In India women take it as a contraceptive after childbirth. Ayurvedic medicine considers cinnamon an astringent, digestive stimulant and aid, carminative, antiseptic, antifungal, antiviral, and blood purifier.

Cinnamon bark is an indispensable ingredient in the kitchen and continues to enjoy popularity as a preferred flavoring agent in many beverages, curried dishes, and meat and rice preparations. The bark is often chewed to protect teeth and gums in Indian homes. Studies show that extracts of cinnamon are excellent antioxidants, activating antioxidant enzymes (Dhuley 1999), and decrease blood sugar levels, providing benefit against diabetic conditions (Onderoglu et al. 1999). Extracts of cinnamon bark also showed antifungal (El-Sayed and Agwa 2001; Inouye et al. 2000; Singh et al. 1995), antimicrobial (Hsieh et al. 2001), and antibacterial (Inouye et al. 2001) properties. When used as a constituent of a traditional herbal formula, cinnamon effectively decreased lung cancer progression (Kamei et al. 2000).

In 2005 global cinnamon production was 134,410 Mt and the area harvested was 175,475 ha (FAOSTAT 2006).

CLOVE

The clove tree (*Szygium aromaticum, Eugenia caryophyllus;* Myrtaceae) is 3.6 to 9.1 m (12-30 ft) tall; its leaves are 7.5 to 15.5 cm (3-6 in) long, 2.5 to 5 cm (1-2 in) wide, and very aromatic. The inflorescence is a terminal branching cyme of 3 to 20 hermaphrodite pale yellow florets, 2.5 to 5 cm (1-2 in) long and borne in clusters. Each flower has a thick cylindrical ovary, above which are four fleshy ovate sepals, and four tiny petals, numerous slender white filaments, and a slender central style. The fruit usually contains one seed. The ovary and sepals constitute the specific part marketed commercially as cloves.

Clove is one of the most valuable ancient Asian spices in international trade. It originated in the Moluccas, the so-called Spice Islands of Indonesia. Indonesia is the world's biggest grower of clove, accounting for more than 50% of world production. Other major growers are Tanzania, Madagascar, Sri Lanka, India, and Malaysia. The spice is often used in tea, *pan,* desserts, and curried dishes of meat, vegetables, and rice for imparting flavor as well as preventing microbial spoilage of food. Koreans use the dried flower buds for the management of asthma and various allergic disorders by oral administration. The clove and clove oil are used in Asian traditional medicine to treat nausea, vomiting, fevers, and dental caries; for toothaches; and externally as an antiseptic to treat wounds and to ease rheumatism.

Clove is generally used all over the world in food preparations and in perfumery and pharmaceutical industries. It is used in India in food preparations, for flavoring, and in spice mixes (*garam masala, pan masala*) as well as in Ayurvedic medicinal preparations. Ripe clove fruits are an essential part in Chinese five-spice powder and curry powders. Clove oil, distilled from the plant, is used in perfumes, medicines, artificial vanillin, and dentifrices.

Research studies demonstrated clove oil's antimicrobial activities against more than 25 different bacteria (De et al. 1999; Dorman and Deans 2000; Hammer et al. 1999), and it has antioxidant (Lee and Shibamoto 2001; Na et al. 2001), anesthetic (Ghelardini et al. 2001), and antifungal (Martini et al. 1996) properties. It was also effective as an antiviral against hepatitis B (Ghazi et al. 2000). Cloves are also used as antimicrobial agents against dental caries and periodontal disease (Cai and Wu 1996).

In 2005 global clove production was 145,370 Mt and the area harvested was 454,110 ha (FAOSTAT 2006).

CURRY LEAF

The curry leaf (*Murraya koenijii* L. Rutaceae) tree is native to southern Asia, specifically southern India. The leaf of this plant is known as the curry leaf because its aromatic leaves are used to flavor curried dishes. Curry leaf is cultivated all over India and in Indonesia, Sri Lanka, China, Vietnam, Jamaica, and in some parts of the United States and Australia. Curry leaf is sometimes referred to as "sweet neem," because the leaves closely resemble neem leaves in their shape and size, though a darker shade of green. The curry leaf tree grows to 4 to 6 m (13-20 ft) and bears abundant leaves throughout the year. The tree branches profusely, produces a number of alternate compound imparipinnate leaves with about 9 to 13 aromatic leaflets, and bears small white flowers and black seeds.

Often cooking starts in a southern Indian household with the quick frying of mustard seeds and curry leaves in shallow hot oil. Fried curry leaves are also used at the final stages of cooking in several dishes and chutneys as garnish. Fresh leaves are extracted to provide the essential oils that are used as a fixative in the soap and perfume industries. Ayurvedic medicine uses the leaves, bark, and roots for enhancing blood circulation, digestion, and metabolism as well as for their anti-inflammatory actions. Curry leaf mixed with buttermilk is used to treat amebiasis, diabetes, and hepatitis. The green leaves are eaten raw as a treatment for dysentery (Drury 1978), and the pulp of the bark relieves bites from poisonous animals (Dastur 1970).

Curry leaf is a rich source of beta-carotene (Bhavani and Kamini 1998), phosphorus, calcium, potassium, iron, and vitamin C (Aykroyd 1966; Joseph and Peter 1985; Reddy and Reddy 1996). Antioxidant vitamins in curry leaves (per gram fresh weight) are lutein, 9,744 ng; alpha-tocopherol, 212 ng; and beta-carotene, 183 ng (Palaniswamy et al. 2003).

Curry leaf possesses hypoglycemic, hypolipidemic, anticarcinogenic, and antioxidant properties (Khan et al. 1995, 1996a,b, 1997). The properties of curry leaf constituents for preventing and controlling dental caries and bad breath have been extensively studied by the Japanese (Inahata et al. 1996, 1997; Tanaka et al. 1998). Some report antioxidant activities of curry leaf extracts (Chakrabarty et al. 2005; Rao et al. 2007) and reduction in blood cholesterol, glucose level, and body weight in animal models (Xie et al. 2006).

Stem bark extractions have been reported to show anti-inflammatory and antimicrobial properties (Ravindra and Srivastava 1992). Other studies show that curry leaf extractions are antibacterial against species of *Staphylococcus* (Akerele and Ayinde 1998) and possess cytotoxic effects (Nutan et al. 1998). The chemical compounds murrayanine, girinimbine, and mahanimbine, isolated from the stem bark of curry leaf tree, showed antifungal activities against several species of human pathogenic fungi (Das et al. 1965).

DILL

Dill (*Anethum graveolens;* Apiaceae) is native to southwestern and central Asia, Mediterranean countries, and Africa. Dill is an essential spice that is also a medicinal, culinary spice, and carminative. Dill is cultivated in Germany, India, western Asia, Rumania, England, and the United States. Dill is a biennial erect plant growing 1 to 1.5 m (3-5 ft) tall with finely dissected, lacy, green foliage characteristic of most Apiaceae members. The flowers are yellow and borne in large, rounded, compound umbels on stiff, hollow stems. The whole inflorescence is 25 to 37 cm (10-15 in) wide. The fruit is a flattened pod. The whole plant is aromatic. The young leaves and the fully developed green fruit are used as a flavoring. The fruits may be dried in shade and stored in closed containers for winter use, but the leaves are normally not dried for storage because they lose their flavor when dried. The fresh leaves are used alone or in combination with other leafy spices, in salad oils, for fried meats and fish, and in sandwiches. The fruiting umbels are chopped into salads and used for flavoring pickles. In Europe dill is used in pickles, salads, and sauerkraut and to flavor vinegars and sauces for fish. Dill is almost indispensable and often called the national spice in Scandinavian countries, because of its extensive use in a number of dishes. Germans, Russians, and Scandinavians tend to use dill mostly for seafood. In India, however, dill seeds are used more than the fresh leaves to flavor lentil and bean dishes and as an essential ingredient in curry. The fresh leaves are used to flavor rice dishes, dips, and deep-fried dishes.

Ayurvedic medicine considers dill seeds to be an excellent aromatic stimulant, tonic, cathartic, and carminative, and as soothing and having a warming effect. The dried seeds are steeped in warm water and taken as a home remedy for stomach problems, nausea, and vomiting. Other medicinal properties attributed to dill in Asian traditional medicines include as a digestive, diuretic, laxative, narcotic, sedative, stimulant, and stomachic. Dill seeds and fresh leaves are used in Asian traditional medicine to treat bruises, cough, dropsy, dyspepsia, flatulence, jaundice, sclerosis, scurvy, mouth sores, stomachache and other stomach ailments, and tumors.

All parts of the dill plant are strongly aromatic. Dill seeds are used whole or ground as a condiment for flavoring curries, meats, sauces, stews, breads, vinegars, pastries, pickles, and vegetable dishes. Fresh leaves are also used as a vegetable and for flavoring the sauces, stews, and other main dishes. Dill oil is used in cosmetic and perfume industries. Dill seeds are used to flavor beverages and drinks, including herbal tea.

Inclusion of dill in the diet resulted in lower serum triacylglycerides, total cholesterol levels (Yazdanparast and Alavi 2001), and blood sugar levels (Essway et al. 1995). The essential oil components carvone and dillapiol derived from dill are reportedly potent antibiotics with demonstrated inhibitory effects against multidrug-resistant bacteria such as *S. aureus* (Chaurasia and Jain 1978; Lee et al. 1999). Other reports of dill's medicinal properties include as an antiproliferative against cancer cell lines, a possible chemopreventive against cancers (Nakano et al. 1998; Zheng et al. 1992), an antimicrobial against yeast and lactic acid bacteria (Shcherbanovsky and Kapelev 1975), and an antifungal (Chaurasia and Kher 1978), indicating its potential use as a food preservative.

FENNEL

Fennel (*Foeniculum vulgare;* Apiaceae) is of Mediterranean, European, and Asian origin. Fennel is cultivated in many parts of Europe, North America, Asia, Egypt, and the Mediterranean. It grows to 0.3 to 0.6 m (1-2 ft) and prefers rich soil and full sun. The leaves are soft, feathery or lacy, bipinnately compound, and further divided into threadlike divisions. The small flowers are pale yellow, have five petals, and are arranged like a dome in compound umbels as is characteristic of the umbelliferous plants. The fruits are small, brown, and resemble tiny ribbed footballs. The seeds are tightly twined in pairs with a conspicuous joint. The fruits of fennel resemble cumin seeds but are larger, a lighter color, and sweeter. Indeed, it is not surprising that all spices of the Apiaceae family (anise, cumin, dill, and caraway) resemble one another in shape and fragrance. Fresh, dried or fried fennel seeds are chewed after meals in Asian cultures as a breath sweetener and to aid digestion. In India it is also used to flavor other masticatory items such as *pan supari* and as a condiment in desserts. The whole plant is edible and is used as a fresh salad, cooked vegetable, spice, and condiment. The seeds are used whole or ground in several meat, vegetable, and rice dishes as a base for sauce or as a spice to impart a sweetish flavor. The dried plants are kept in grain storage areas to serve as an insect repellent.

Fennel has a long history of herbal use and is a commonly used household remedy for treating problems related to the digestive system. A tea made from fennel seeds is a common treatment for indigestion. Fennel is described in Asian as well as European traditional medicines as soothing and is used as an analgesic, anti-inflammatory, antispasmodic, aromatic, carminative, diuretic, expectorant, laxative, stimulant, and stomachic. Ayurvedic medicine believes fennel seeds have a heating property and thus prohibits its use by pregnant women during early pregnancy to avoid a miscarriage. An infusion of the seeds is used to treat indigestion, abdominal distension, and stomach pains, and to expel gas in children and adults. An essential oil extracted from the fully ripened and dried seed is used for medicinal purposes and also in perfumery.

Scientific studies have documented fennel's ability to improve digestion by significantly stimulating intestinal enzymes, particularly lipase and amylase, contributing to fennel's well-recognized digestive stimulant action in traditional use (Platel and Srinivasan 2001). The essential oils extracted from fennel are antimicrobial (Elgayyar et al. 2001) and active against 25 genera of bacteria, including animal and plant pathogens, food-poisoning and spoilage bacteria, demonstrating a broad range of antioxidant and antibacterial activity (Ruberto et al. 2000). Fennel's essential oils are hypoglycemic (Essway et al. 1995). Fennel leaf and seed extracts are antioxidants (Nakayama et al. 1996; Terasawa et al. 2001), antiviral, and inhibitory against hepatitis C virus (Hussein et al. 2000). A number of studies also explored fennel's use in the food industry because of its broad range of antimicrobial and antibiotic activities (Guillen and Manzanos 1996; Hammer et al. 1999; Piccaglia et al. 1993; Tkachenko et al. 1999). Some controlled studies also report an estrogenic property of fennel seeds when fed to rats and mice (Annusuya et al. 1988; Malini et al. 1985), which may explain and validate the traditional use of fennel to increase milk secretion in feeding mothers, promote menstruation in women, facilitate birth during delivery of a child, alleviate symptoms of the male climacteric, and increase libido in males (Albert-Puleo 1980).

GINGER

Ginger (*Zinziber officinale;* Zingiberaceae) is a plant native to Southeast Asia. For more than 5,000 years ginger has been prized as much for its medicinal properties as for its unique flavor and culinary value. References to ginger as a medicine and spice appear in ancient Chinese and Sanskrit writings; in Persian, Arabic, Greek, and Roman texts; and in medieval writings. Ginger is grown widely throughout tropical Asia, Japan, the West Indies, South America, and Western Africa. India is the world's largest ginger producer. Ginger is a perennial plant with an erect stem growing 0.6 to 1.2 m (2-4 ft) high and bearing narrow, long, lanceolate leaves that die annually. The flowering stalk rises directly from the rhizome, ending in an oblong scallop spike. Each spike bears aromatic white or yellowish green flowers streaked with purple. The rhizome is branched and the individual divisions are often referred to as fingers or hands.

Ginger is normally available in its fresh form; in its processed dried, black, or white form; or preserved in syrup. Dried ginger is made by washing and drying the rhizomes, while black ginger is prepared by scalding the rhizomes with water and drying them. In making white ginger, the outer layers of the rhizomes are peeled off before washing and drying. Preserved ginger is made by peeling the rhizomes and boiling them in syrup.

Ginger is an important component of condiments, curries, pickles, flavored drinks, and syrup. Fresh crushed ginger is added to beverages and candies. Ginger has been popular since the 1500s in the West as a spice in gingerbread and ginger ale. It is a tangy spice used all over the world in baking, cooking, and flavoring beverages. For more than 2,500 years, ginger has also played an important role in Asian medicine as a folk remedy to promote cleansing of the body, stimulate the appetite, cure motion sickness, and prevent morning sickness in pregnant women, and as a tonic.

In Asian medical systems ginger is considered hot and warming and is often used to decrease coolness or increase heat in the body. Ayurvedic medicine uses ginger for digestive ailments, headaches, anorexia, diabetes, cholera, constipation, rheumatism, and liver inflammation (Aiyer and Kolammal 1966). Traditional Chinese medicine practitioners have used ginger either alone or in combination with other herbs for more than two centuries primarily for illness related to the gastrointestinal tract and inflammatory joint conditions. Ginger is frequently added to other herbal compounds to aid in digestion and increase the action of other herbs.

A number of chemical compounds have been isolated from ginger, including gingerol, a ginger oleoresin (combination of volatile oils and resin) that accounts for its characteristic aroma and therapeutic properties. Components of gingerol (zingiberone, bisabolene, camphene, geranial, linalool, and borneol) have been found to possess beneficial properties for treatment of poor digestion, heartburn, and vomiting and prevention of motion sickness. In addition, ginger is a stimulant and carminative (Ernst and Pittler 2000). Other conditions for which ginger is used in Asian medical systems include atherosclerosis, migraine headaches, and rheumatoid arthritis, and it is used to improve blood circulation and as an effective gargle for sore throats.

Ginger provides potential health benefits by inhibiting lipid peroxidation (Shobana and Naidu 2000), lowering blood cholesterol and sugar levels (Giri et al. 1984; Singhal and Joshi 1983), and acting as a digestive stimulant (Platel and Srinivasan 2000). Long-term dietary intake of ginger has a hypoglycemic and hypolipidemic effect (Ahmed and Sharma 1997). Ginger is also an effective antioxidant (Ahmed et al. 2000), and the compounds in ginger have the ability to suppress proliferation of human cancer cells by inducing apoptosis (Surh 1999).

LEMONGRASS

Lemongrass (*Cymbopogon citratus;* Poaceae) is native to South Asia, Southeast Asia, and Australia. The genus *Cymbopogon* has more than 50 species, all of which are important as a culinary herb, spice, and medicine. The common name lemongrass is indicative of its grasslike leaves and lemonlike scent. Lemongrass is cultivated in tropical Asian countries, including India, Sri Lanka, Thailand, Vietnam, Cambodia, and the Indonesian islands as a culinary and medicinal herb. It is a perennial grass that reaches 1.2 to1.8 m (4-6 ft). It grows in bushy clumps and has slender leaves and grasslike flowers in terminal spikes or racemes. The leaves are pale green, long and narrow, and similar to most grass species. They are sessile, simple, and glabrous and have parallel venation with an acuminate tip and sheathing base. The lemongrass plant does not have true stems. The leaf sheath is tubular and serves as a pseudostem. The flowers are bisexual, bearing male and female flowers on the same plant. The inflorescence is a raceme. Lemongrass is also grown as a desirable ornamental grass for its long, straplike leaves that have gracefully drooping tips. It is a popular culinary herb added to curry pastes and to pork, poultry, fish, seafood, and vegetable dishes. The leaves are steam distilled to extract lemongrass oil that is used in the perfume industry and in making soaps, cosmetics, and scented skin creams.

Lemongrass is widely used in Chinese and Indian medicine for numerous ailments. Lemongrass oil is well known as an antiseptic used to treat athlete's foot. Medicinally the essential oil is used to treat acne, as a carminative, and as an insect repellent. In Suriname's traditional medicine, lemongrass treats headaches, coughs, cuts, asthma, and bladder disorders and is used as a diaphoretic. The leaves are also used to make herbal teas and as a flavoring in cakes and other baked goods. One of the active ingredients in lemongrass, citral, is known to aid digestion and relieve spasms, muscle cramps, rheumatism, and headaches. It is a home remedy for the treatment of flu, fevers, headaches, diarrhea, and stomach problems and is used as a diuretic. Citral is used as an antiseptic to deodorize clothing and footwear. The leaves are used in Cuban folk medicine as an antihypertensive and anti-inflammatory. Tea from the leaves is used as a home remedy in Brazil as a spasmolytic, analgesic, anti-inflammatory, antipyretic, diuretic, and tranquilizer.

Lemongrass is becoming a popular herb in the kitchen and a favorite of gardeners in North America. It is a versatile herb used in teas, beverages, herbal medicines, and Asian cuisine. The leaves and stems of the young plant are reported to have antifungal, insect repellent, and bacteriostatic activities (Ponglux et al. 1987). Aqueous extracts of lemongrass demonstrated mitodepressive effects (Williams and Omoh 1996), antinociceptive properties (Viana et al. 2000), and antitumor activities against skin cancer (Connor 1991). Decoctions of lemongrass leaves demonstrated diuretic and anti-inflammatory (Carbajal et al. 1989) properties. Researchers have reported antibacterial (Onawunmi et al. 1984), antitumor (Murakami et al. 1994), antihepatocarcinogenic (Puatanachokchai et al. 2002), and antimutagenic (Vinitketkumnuen et al. 1996) properties.

LICORICE

Licorice (*Glycyrrhiza glabra;* Fabaceae) is native to China and India. Licorice roots and leaves were used in food and medicine as early as ancient Egyptian times, since the third century BC. It is one of the essential components of the popular Chinese five-spice powder. Licorice is grown commercially for its roots and is widely cultivated as a crop in southern Europe, Asia, and South America. The licorice plant grows to 1.2 m (4 ft) and spreads to approximately 0.9 m (3 ft). The erect stem bears greenish blue leaves that are divided into oval leaflets. The flowers are small, white with a pale purple blue tinge, and on erect stems. The fruits are brown pods covered with hooked prickles. Its roots have a pleasant sweet taste.

In Chinese herbal medicine, licorice is an ingredient in almost all herbal preparations and is used as a harmonizing adaptogen to enhance the effects of other herbs in the formula. In Ayurvedic medicine, licorice is called the *yashti madhu,* the "sweet stick," and is used as an anti-inflammatory agent and aphrodisiac, and to improve eyesight and general body strength. Licorice is often used to help cure the common cold, coughs, and sore throats and as a flavoring agent.

The cut, dried roots are used in expectorants and as antitussive agents in cough syrups and cough drops. Research over the last two decades has revealed that licorice is an herb with excellent hepatoprotective properties. It is an antioxidant and antiviral, enhances interferon and antibody production, promotes T cell activity in the liver, and protects the liver from autoimmune liver injury (Konovalova 2000; Luper 1999; Vaya et al. 1997).

Licorice can also benefit patients receiving chemotherapy for cancer, preventing the drugs' hepatic side effects. In Japan glycyrrhizin, the main active component in licorice, is now a standard medical treatment for liver problems, including hepatitis B virus infection. Glycyrrhizin has anti-inflammatory action and may inhibit breakdown of cortisol (Soma et al. 1994; Whorwood et al. 1993).

The flavonoids in licorice help heal digestive tract cells and act as antioxidants protecting liver cells. The flavonoids have been shown to kill *Helicobacter pylori,* the bacteria that cause most ulcers and stomach inflammation (Beil et al. 1995). An extract of licorice, called liquiritin, is effective for treating melasma, a pigmentation disorder of the skin (Amer and Metwalli 2000), and preventing skin cancer (Agarwal et al. 1991; Jung et al. 2001; Razina et al. 2000; Shankel and Clarke 1990; Wang 1994). Licorice is antimicrobial (Fukui et al. 1988), antiviral (Hussein et al. 2000; Yamashiki et al. 1999), anti-ulcerogenic (Khayyal et al. 2001), and hypolipidemic and prevents aflatoxin formation (El-Shayeb and Mabrouk 1984), atherosclerosis (Khushbaktova 1991), and dental caries (Namba et al. 1985). Licorice root is a highly regarded remedy in the major herbal traditions of the world. Although mainly considered a hormonal, digestive, or respiratory remedy, licorice is now emerging as a quality liver botanical.

NUTMEG

Nutmeg (*Myristica fragrans;* Myristicaceae) is endemic to the Indonesian Banda Islands (Moluccas). Nutmeg has been used in China since the seventh century and in India since the ninth century for stomach problems and mouth sores. Nutmeg is cultivated in India, Iran, the West Indies, Indonesian islands, and South America. Nutmeg is a tropical evergreen tree growing to 20.3 m (70 ft). The trunk is grayish brown with smooth bark. The leaves are brownish, oblong or elliptical, glabrous, dark green, and glossy above, with a paler green on the lower surface. The flowers are small, light yellow, and bell shaped. The male and female flowers are borne on separate trees (dioecious). The fruit is a drupe, grooved like an apricot, and reddish to yellowish. The fruits turn a dark brown, crinkle up, and split into halves on maturity. The brown seed (the nutmeg) is surrounded by a scarlet aril (the mace). The pendulous nutmeg fruit gives the tree an attractive appearance. The seeds are collected when the fruits ripen and split. The scarlet aril surrounding the seed is separated and dried in the sun. The seed is dried in the sun for two to three months, turned every few days for uniform drying, until the kernel starts to rattle within the shell, and then the shell is cracked and the seed removed.

The unripe pericarp of the nutmeg fruit is preserved in sugar, salted and dried as a condiment, or made into jellies and other preserves. The pulp of the ripe fruit is also eaten. Grated or ground nutmeg is used as a spice and a warm aromatic, slightly bitter in taste. It is often added to desserts, custards, confectionery, meats, sausages, sauces, bakery, puddings, pies, certain vegetable dishes, and milk drinks such as flavored coffee and eggnog. The mace is also popular for its strong aroma.

Nutmeg is used for both culinary and medicinal purposes in its area of origin, primarily for diarrhea, mouth sores, and insomnia. The Arabs used nutmeg for stomach problems and as an aphrodisiac, while the Indians used it as a spice and described it as being stimulating, and aiding digestion. Ayurvedic and Unani medicine use it to ease headaches and fever. The mace is made into an ointment and used for rheumatism. Nutmeg is one of several herbs kept in home medicine kits in India and is given to children to induce sleep and to cool excessive heat in the body.

Ayurvedic medicine prescribes it for headaches, fevers, mouth sores, excessive heat of the body, foul breath, and intestinal weakness and as an analgesic and a sedative. Despite its perception as a narcotic in the West, very few cases of nutmeg's hallucinogenic or other psychoactive properties other than a mild sedative action are reported in the countries of origin.

Nutmeg is used in small dosages in traditional medicine to treat flatulence, diarrhea, vomiting, and nausea; aid digestion; and improve the appetite (Van and Cox 1994). The reported functional, medicinal, and narcotic properties of nutmeg are due to myristicin. Myristicin when taken in large dosages can cause hallucinations, vomiting, and epileptic seizures. Nutmeg has been reported to fight dental plaque by inhibiting the cariogenic bacterium, *S. mutans* (Namba et al. 1985).

Nutmeg oil (an essential oil extracted from the nut) also showed effective antibacterial activity against 25 different genera of bacteria, including animal and plant pathogens and food-poisoning and spoilage bacteria (Dorman and Deans 2000), and has antifungal activity against several pathogenic fungi, including *A. flavus* and *Candida albicans* (Rahman et al. 2000). The seed extracts of nutmeg also demonstrated hypercholesterolemic activity; reduced serum and LDL cholesterol; prevented the accumulation of cholesterol, phospholipids, and triglycerides in the liver, heart, and aorta; and dissolved atheromatous plaques of the aorta in rabbits (Sharma et al. 1995). The medicinal property of mace and its potential use as a possible chemopreventive, anti-inflammatory, analgesic, and antithrombotic agent have been studied in rodents (Jannu et al. 1991; Nakajima 1989;

Olajide et al. 1999; Ozaki et al. 1989). Topical application of mace also reduced symptoms of skin cancer in mice (Jannu et al. 1991).

SAFFRON

Saffron (*Crocus sativus;* Iridaceae) is a native of Eurasia known as a spice, food colorant, and medicine since ancient times, dating back to 3000 BC in Sumerian texts and as early as 500 BC in India. It is the world's most expensive spice and is derived from the dried stigmas of its flower. Spain and India are the major producers of saffron. Saffron is also grown in Egypt, Iran, and New Zealand. Traditional uses for saffron are mainly as a spice and a medicine. A pinch of saffron is added as a food colorant and flavoring agent to desserts, rice dishes, milk shakes, and beverages. It is an important ingredient of Ayurveda and Unani medical systems in India.

Saffron is a perennial plant grown in home gardens and also cultivated commercially for its spice. It grows 10 to 30 cm tall and has long narrow leaves. The leaves are radical, linear, dark green above, pale green below, and enclosed in a membranous sheath. From the bulb rises a long, white, slender flower stalk that bears the large, beautiful, lilac to mauve flower. The flowers have six petals, and long styles hang out of the flowers. The three stigmas inside the flower are large and bright orange red. The stigmas of saffron are the part used in food and medicine and have a slightly bitter taste. It is proposed to have sedative, antispasmodic, expectorant, and aphrodisiac properties. A common practice in affluent Indian families is to mix a pinch of saffron with milk and drink it at bedtime for a sound sleep. Pregnant women in India drink this mix every day for blood purification, and it is a common belief that intake of saffron during pregnancy will produce a baby with fair skin and good complexion.

Chemical analysis of saffron extracts revealed that its characteristic compounds include the carotenoids crocin and crocetin and the monoterpene aldehydes picrocrocin and safranal. Saffron is used as a culinary seasoning and coloring. Saffron's chemistry and its functional properties have been examined by several researchers in the past decade. The flavor of saffron and its attributed benefits in traditional medicine have attracted the interest of several research groups (Winterhalter and Straubinger 2000).

It has been demonstrated that saffron taken with milk is a potential antioxidant, reducing lipoprotein oxidation susceptibility and preventing oxidative stress (Verma and Bordia 1998). Saffron is also effective in the treatment of acne and inflammation (Jampani et al. 2001), and it is described as a suitable health food for people with dementia (Sugiura 1999). Saffron extracts and its active constituents, crocetin and crocin, have been demonstrated to enhance memory and are suggested as useful in treating neurodegenerative disorders (Abe and Saito 2000; Kazuho and Hiroshi 2000). Because of its cytotoxic effect, it is suggested for treatment of colon cancer (Escribano et al. 2000; Garcia-Olmo et al. 1999). Other studies showed that crocin significantly increased blood flow in the retina, which improves oxygenation and nutrient supply to retinal structures, facilitates retinal-function recovery, and indicates a potential use to treat ischemic retinopathy or age-related macular degeneration (Xuan et al. 1999). Antitumor properties of saffron have been demonstrated by several researchers over the past decade (Abdullaev and Frenkel 1992a,b; Aruna and Sivaramakrishnan 1990; Salomi et al. 1991). Pharmacological studies have demonstrated that saffron extracts or its constituents have antitumor (Nair et al. 1991) and hypolipidemic effects as well as radical-scavenging properties (Ríos et al. 1998). Chemicals isolated from the corms of saffron plants also exhibited antitumor activity (Escribano et al. 1999, 2000).

Antinociceptive, anti-inflammatory (Hosseinzadeh and Younesi 2002), anticonvulsant (Hosseinzadeh and Khosravan 2002), and antidepressant effects have also been reported in animal studies (Hosseinzadeh et al. 2004) and human trials (Akhondzadeh et al. 2004). Other studies have shown that saffron protects platelets from aggregation and prevents lipid peroxidation (Jessie and Krishnakantha 2005). Some researchers suggest a potential role for saffron as an antigenotoxic, antioxidant, and chemopreventive agent that could be used as an adjuvant in chemotherapeutic applications (Premkumar et al. 2006).

STAR ANISE

Star anise (*Illicium verum;* Illiaceae) originated in southern China and Vietnam. Star anise is grown extensively in China, Laos, Vietnam, the Philippines, and Jamaica. It is the fruit of a small evergreen tree that grows to about 10 m (30 ft) The leaves are green and broad, alternate or opposite. The yellow flowers of the tree produce dark-brown fruits that are star shaped with eight segments. Each segment has a pea-sized seed, which ripens and then opens into a star shape. Both the pod and seed are used as the spice. The spice has a slightly sweet licorice flavor. Its flavor closely resembles anise, and its name reflects this property as well as the eight-pointed star-shaped pods that are characteristic to this plant species and form the commercial spice. It is an essential spice and a flavoring agent in several Asian cuisines, and it is often chewed after meals to promote digestion and sweeten the breath. It is available commercially as whole pods, broken pods, or finely ground powder. Star anise is used as a spice by itself and as a component of other spices such as the Chinese five-spice powder, Indian *biriyani* mix, *pulao* mix, and other mixtures used to flavor curried dishes and rice preparations. Star anise is important in Pakistani, Persian, Arabic, Indonesian, and other Asian and Western cuisine and is often used to flavor wine, desserts, tea, and liqueurs. The dried fruit and the seeds are both medicinal, as is the oil extracted from the fruit.

Asian traditional medicine proposes that it fights cough and bronchitis by loosening phlegm. It also has an antispasmodic effect on the intestines and is thus used to relieve stomach cramps. It is a prescribed remedy for appetite loss, bronchitis, and cough. In Asian medicine, star anise is used for indigestion, arthritis, and intestinal cramps; to enhance relaxation and emotional balance; and as an aphrodisiac, antiseptic, carminative, expectorant, insect repellent, and stimulant. Star anise pods are also used in aromatherapy, incense, and perfumery.

Currently several preparations of beverages, including herbal teas, use star anise as a flavoring agent. It is widely used in pickles and preservatives to impart a desirable flavor. Star anise is an important ingredient in Chinese medicine for a number of ailments (Zhang et al. 1999). In Japan and in the West, several patents exist that use star anise in dentifrices, food, and feed, to prevent formation of dental plaque and caries (Oshima and Mitsunaga 1999). It is reported as an analgesic (Nakamura et al. 1996) and as a high-quality, low-priced, and rich natural antioxidant (Shang et al. 1994). It is also used in insecticides, microbicides, and air fresheners (Ho et al. 1997; Matsumura et al. 1994; Miyazawa et al. 1993).

SWEET BASIL

Sweet basil (*Ocimum basilicum* L.; Lamiaceae, or Labiatae) is a close relative of India's most popular and sacred Holy Basil, *Tulsi* (*O. sanctum*). There are 50 or 60 different species of basil. It is native to Southeast Asia and central and northeastern Africa. Basil has been cultivated as a culinary herb around the world since ancient times. It is cultivated extensively in Asia and in parts of Europe, including France, Hungary, Greece, the Middle East, and the United States. Basil is a small, bushy, erect annual or short-term perennial. The leaves are simple, opposite, pubescent or glabrous, and light green on the top and gray green on the underside. They grow 5 to 10 cm (2-4 in) long. The lavender white or purplish flowers (spikes) yield numerous seeds. Basil leaves and flowers have a strong, fresh, clovelike scent. Today basil is one of the fastest growing herbs in terms of usage in North America. Because of its popularity there and in Europe, a number of new varieties with variations in color, leaf size, flavor, and fragrance have been developed, and they are often listed as an herb or a medicinal in most seed catalogs.

Basil is cultivated for local markets to sell fresh as a culinary herb; dried as a condiment or spice; as a source of aromatic essential oil for use in foods, flavors, and fragrances; and as a potted herb and bedding plant. Traditionally, in Asian medical systems, basil has been used as a medicinal plant in the treatment of headaches, coughs, intestinal worms, and kidney problems. Its use as an insect repellent and an antibacterial agent is also very common in Asian cultures.

Basil leaves, fresh or dried, are used in cooking and flavoring dishes. Basil is used in cooking most often with garlic, tomatoes, eggplant, and Italian dishes, and to flavor vinegar, condiments, liqueurs, and oil. Basil is used to scent soaps and perfumes. The most common basil types available in the market include California basil with a uniform particle size, pale green color, and sweet flavor; French basil, which is slightly sweeter and darker in color; and Egyptian basil, which has a camphorlike fragrance and a strong minty flavor.

Reportedly, inhaling the essential oil refreshes the mind and stimulates a sense of enchantment. The leaf extract is considered a tonic and aphrodisiac, stimulating the adrenal cortex. The leaves are used as a mosquito repellent and to treat ringworm, insect bites, snakebites, and acne. Basil leaf extract is used in massage oils as a nerve tonic to ease tired and overworked muscles.

A paste of basil root is also used as an antiseptic in treating scorpion and snakebites in Ayurvedic medicine. Currently, *O. basilicum* and *O. sanctum* oils are being studied for their anti-inflammatory and antiulcerogenic activity (Singh 1999; Singh and Majumdar 1999). Basil oils also have antimicrobial effects and may be a potential preservative in food preparations (Lachowicz et al. 1998). Basil may also prove to be a valuable source of anticarcinogenic agents (Aruna and Sivaramakrishnan 1992). Basil leaf extracts are antidiarrheal (Ilori et al. 1996) and possess blood-sugar-lowering agents (Chattopadhyay 1999). In addition, Holy Basil may have the ability to prevent the early events of carcinogenesis (Karthikeyan et al. 1999). Other studies provide evidence of potent anti-HIV-1 (Yamasaki et al. 1998), antioxidant (Hinneburg et al. 2006; Maulik et al. 1997), hypoglycemic, and hypolipidemic (Rai et al. 1997; Vats et al. 2002) activities of basil leaves and leaf extracts.

TURMERIC

Turmeric (*Curcuma longa;* Zingiberaceae) is a native of tropical Asia, specifically India. It is not surprising that sometimes turmeric is also called Indian saffron because of its brilliant yellow color. The value of turmeric as a food and medicine has been recognized since the early ages. It is reported that Marco Polo described turmeric as "a vegetable with the properties of saffron" in 1280. It has been used as a dye, medicine, and flavor agent since 600 BC. India is the world's primary producer of turmeric. It is also grown in China, Taiwan, Japan, Burma, Indonesia, and throughout Africa.

Turmeric is a perennial shrub that can grow to about 1 m (3.3 ft), with oblong rhizomes that are deep orange inside. Mature leaves are about 60 cm (2 ft) long, lanceolate, with a petiole, tapered at each end, smooth, and of a uniform green. The flowers are cone shaped, 10 to 15 cm (0.3-0.5 ft) long, pale yellow, in bunches of three or five, and surrounded by bracteoles. The rhizomes are dug up and separated into mother and finger rhizomes. The rhizomes are cured by boiling, steaming, or cooking in water and then drying and polishing or peeling the dried tissue.

Turmeric has a distinct fragrance and a slightly bitter taste. It has been used medicinally throughout Asia to treat stomach and liver ailments. It was also used externally, to heal sores and as a cosmetic. Indonesians use turmeric to color their bodies as part of their wedding ritual. In India, since ancient times, it is most popular as a food-coloring agent or as a condiment that is used in almost every dish. It is also used as an age-old beauty treatment by Indian women for washing the body and face. Regular topical application of turmeric from a very young age is believed to reduce hair growth in women and ensure smooth clear skin.

Turmeric is commercially available as fresh or dried rhizome or as powder in Asian grocery stores and as encapsulated powder in health-food stores and pharmacies. It is an essential ingredient of curry powder, and it is used extensively in Indian dishes and in almost all Southeast Asian cooking. Turmeric is also added to pickles and relishes and is occasionally used in place of saffron to provide color and flavor.

Its modern scientifically approved applications originate from its traditional uses in Asia to treat coughs, sore throats, open wounds, and stomach ulcers, and as an anti-inflammatory agent, a digestive aid, and a blood purifier. It also helps regulate disordered functions of the body. Turmeric is used extensively in the traditional Indian medical systems (Ayurveda, Unani, and Siddha) and is officially entered in the *Ayurvedic Pharmacopoeia of India* (Government of India 1989). In both the Ayurvedic and Siddha systems of medicine, a turmeric paste is used topically as well as internally to treat ulcers and scabies (Charles and Charles 1992). Turmeric also serves as a disinfectant and antiseptic in the treatment of wounds, measles, boils, and other skin disorders. It is also used extensively in traditional Chinese and Japanese medicines: it is in the *Pharmacopoeia of the People's Republic of China* (Tu et al. 1992) and in the *Japanese Herbal Medicines Codex* (Japanese Ministry of Health and Welfare 1993).

A pinch of turmeric boiled in milk is a traditional home remedy in several Asian cultures for colds, sore throats, and jaundice. When mixed with lime, turmeric turns red and hence is used to make chemical test papers that are used to detect basic, or alkaline, conditions. Turmeric contains curcumin and an essential oil (turmerone and zingiberins). Curcumin is a diet-derived promising chemopreventive agent studied for anticarcinogenic and antioxidant properties (Kelloff et al. 2000). Curcumin has anti-inflammatory, antifungal, and cytotoxic effects (Apisariyakul et al. 1995; Arora et al. 1971; Sung et al. 1997); possesses antifungal and antibacterial properties (Roth et al. 1998); has demonstrated the ability to lower cholesterol levels and heal ulcers; and has shown antioxidant, antimutagen, and anticarcinogen properties in numerous studies (Dorai et al. 2000; Graf 2000; Polasa et al. 1992; Selvam et al. 1995; Van Dau et al. 1998). Turmeric is reportedly effective in the treatment of a number of conditions, including arthritis, digestive problems, and obesity. It also protects the liver.

REFERENCES

Abdullaev FI, Frenkel GD. 1992a. Effect of saffron on cell colony formation and cellular nucleic acid and protein synthesis. Biofactors 3(3): 201-204.

Abdullaev FI, Frenkel GD. 1992b. The effect of saffron on intracellular DNA, RNA and protein synthesis in malignant and non-malignant human cells. Biofactors 4(1): 43-45.

Abe K, Saito H. 2000. Effects of saffron extract and its constituent crocin on learning behavior and long-term potentiation. Phytotherapy Research 14: 149-152.

Agarwal R, Wang ZY, Mukhtar H. 1991. Inhibition of mouse skin tumor-initiating activity of DMBA by chronic oral feeding of glycyrrhizin in drinking water. Nutrition and Cancer 15(3-4): 187-193.

Ahmed RS, Seth V, Pasha ST, Banerjee BD. 2000. Influence of dietary ginger (*Zingiber officinales* Rosc) on antioxidant defense system in rat: Comparison with ascorbic acid. Indian Journal of Experimental Biology 38(6): 604-606.

Ahmed RS, Sharma SB. 1997. Biochemical studies on combined effects of garlic (*Allium sativum* Linn) and ginger (*Zingiber officinale* Rosc) in albino rats. Indian Journal of Experimental Biology 35: 841-843.

Aiyer KN, Kolammal M. 1966. Pharmacognosy of ayurvedic drugs. No. 3. Department of Pharmacognosy, University of Kerala, Trivandrum.

Akerele O, Ayinde BA. 1998. Antibacterial activities of the volatile oil and aqueous extract of *Murraya koenigii* leaves. Nigerian Journal of Natural Products and Medicine 2: 44-45.

Akhondzadeh S, Fallah-Pour H, Afkham K, Jamshidi AH, Khalighi-Cigaroudi F. 2004. Comparison of *Crocus sativus* L. and imipramine in the treatment of mild to moderate depression: A pilot double-blind randomized trial [ISRCTN 45683816]. BMC Complementary and Alternative Medicine 4: 12.

Albert-Puleo M. 1980. Fennel and anise as estrogenic agents. Journal of Ethnopharmacology 2(4): 337-344.

Al-Zuhair H, El-Sayeh B, Ameen HA, Al-Shoora H. 1996. Pharmacological studies of cardamom oil in animals. Pharmacological Research 34(1-2): 79-82.

Amer M, Metwalli M. 2000. Topical liquiritin improves melasma. International Journal of Dermatology 39: 299-301.

Annusuya S, Vanithakumari G, Megala N, Devi K, Malini T, et al. 1988. Effect of *Foeniculum vulgare* seed extracts on cervix and vagina of ovariectomised rats. Indian Journal of Medical Research 87: 364-367.

Apisariyakul A, Vanittanakom N, Buddhasukh D. 1995. Antifungal, activity of turmeric oil extracted from *Curcuma longa* (Zingiberaceae). Journal of Ethnopharmacology 30: 163-169.

Arora RB, Basu N, Kapoor V, Jain AP. 1971. Anti-inflammatory studies on *Curcuma longa* (turmeric). Indian Journal of Medical Research 59: 1289-1295.

Aruna K, Sivaramakrishnan VM. 1990. Plant products as protective agents against cancer. Indian Journal of Experimental Biology 28(11): 1008-1011.

Aruna K, Sivaramakrishnan VM. 1992. Anticarcinogenic effects of some Indian plant products. Food and Chemical Toxicology 30(11): 953-956.

Aykroyd WR. 1966. The nutritive value of Indian foods and planning of satisfactory diets. 6th ed. Indian Council of Medical Research, New Delhi. p. 53-58.

Beil W, Birkholz C, Sewing KF. 1995. Effects of flavonoids on parietal cell acid secretion, gastric mucosal prostaglandin production and *Helicobacter pylori* growth. Arzeneimittel-Forschung 45: 697-700.

Bhavani KN, Kamini D. 1998. Development and acceptability of a ready-to-eat beta-carotene rich, maize based supplementary product. Plant Foods for Human Nutrition 52(3): 271-278.

Cai L, Wu CD. 1996. Compounds from *Syzygium aromaticum* possessing growth inhibitory activity against oral pathogens. Journal of Natural Products 59: 987-990.

Carbajal D, Casaco A, Arruzazabala L, Gonzalez R, Tolon Z. 1989. Pharmacological study of *Cymbopogon citrates* leaves. Journal of Ethnopharmacology 25(1): 103-107.

Chakrabarty MM, Mukherjee S, Bagch UB, Hoque Sk A, Sen D. 2005. Studies on the antioxidant activity of oil extract of curry leaf in oil. Journal of the Oil Technologists' Association of India 37(3): 7-9.

Chang TS. 1999. Herbal composition for stimulating blood circulation. U.S. Patent 5942233 A 19990824. p. 6.

Charles V, Charles SX. 1992. The use and efficacy of *Azadirachta indica* ("neem") and *Curcuma longa* ("turmeric") in scabies. A pilot study. Tropical and Geographical Medicine 44: 178-181.

Chattopadhyay RR. 1999. A comparative evaluation of some blood sugar lowering agents of plant origin. Journal of Ethnopharmacology 67(3): 367-372.

Chaurasia SC, Jain PC. 1978. Antibacterial activity of essential oils of four medicinal plants. Indian Journal of Hospital Pharmacy 15(6): 166-168.

Chaurasia SC, Kher A. 1978. Activity of essential oils of three medicinal plants against various pathogenic and nonpathogenic fungi. Indian Journal of Hospital Pharmacy 15(5):139-141.

Connor MJ. 1991. Modulation of tumor promotion in mouse skin by the food additive citral (3,7-dimethyl-2, 6-octadienal). Cancer Letters 56: 25-28.

Das KC, Chakraborty DP, Bose PK. 1965. Antifungal activity of some constituents of *Murraya koenigii* Spreng. Experientia 21(6): 340.

Dastur JF. 1970. Medicinal plants of India and Pakistan. 3rd ed. Taraporevala Sons, Bombay. p. 115.
De M, Krishna DA, Banerjee AB. 1999. Antimicrobial screening of some Indian spices. Phytotherapy Research 13(7): 616-618.
Dhuley JN. 1999. Anti-oxidant effects of cinnamon (*Cinnamomum verum*) bark and greater cardamom (*Amomum subulatum*) seeds in rats fed high fat diet. Indian Journal of Experimental Biology 37(3): 238-242.
Dorai T, Gehani N, Katz A. 2000. Therapeutic potential of curcumin in human prostate cancer. II. Curcumin inhibits tyrosine kinase activity of epidermal growth factor receptor and depletes the protein. Molecular Urology 4: 1-6.
Dorman HJD, Deans SG. 2000. Antimicrobial agents from plants: Antibacterial activity of plant volatile oils. Journal of Applied Microbiology 88(2): 308-316.
Dorman HJD, Surai P, Deans SG. 2000. *In vitro* antioxidant activity of a number of plant essential oils and phytoconstituents. Journal of Essential Oil Research 12(2): 241-248.
Drury HC. 1978. The useful plants of India. 2nd ed. Allen, London. p. 78.
Eigner D, Scholz D. 1990. The magic book of Gyani Dolma. Pharmazie in Unserer Zeit 19(4): 141-152.
Elgayyar M, Draughon FA, Golden DA, Mount JR. 2001. Antimicrobial activity of essential oils from plants against selected pathogenic and saprophytic microorganisms. Journal of Food Protection 64(7): 1019-1024.
El-Sayed TI, Agwa HE. 2001. Effect of some essential spice oils on growth and aflatoxins production by *Aspergillus flavus*. Al-Azhar Journal of Microbiology 51: 129-137.
El-Shayeb NMA, Mabrouk SS. 1984. Utilization of some edible and medicinal plants to inhibit aflatoxin formation. Nutrition Reports International 29(2): 273-282.
Ernst E, Pittler MH. 2000. Efficacy of ginger for nausea and vomiting: A systematic review of randomized clinical trials. British Journal of Anaesthesia 84(3): 367-371.
Escribano J, Diaz-Guerra M, Jose M, Riese HH, Alvarez A, et al. 2000. The cytolytic effect of a glycoconjugate extracted from corms of saffron plant (*Crocus sativus*) on human cell lines in culture. Planta Medica 66(2): 157-162.
Escribano J, Jose M, Diaz-Guerra M, Riese HH, Jesús O, et al. 1999. *In vitro* activation of macrophages by a novel proteoglycan isolated from corms of *Crocus sativus* L. Cancer Letters 144(1): 107-114.
Essway GS, Sobbhy HM, El-Banna HA. 1995. The hypoglycemic effect of volatile oil of some Egyptian plants. Veterinary Medical Journal Giza 43(2): 167-172.
FAOSTAT (Statistics Division of the Food and Agriculture Organization of the United Nations). 2006. http://faostat.fao.org. Accessed February 2006.
Fatehi M, Farifteh F, Fatehi-Hassanabad Z. 2004. Antispasmodic and hypotensive effects of *Ferula* asafoetida gum extract. Journal of Ethnopharmacology 91(2-3): 321-324.
Fukui H, Goto K, Tabata M. 1988. Two antimicrobial flavanones from the leaves of *Glycyrrhiza glabra*. Chemical and Pharmaceutical Bulletin 36(10): 4174-4176.
Garcia-Olmo DC, Riese HH, Escribano J, Ontanon J, Fernandez JA, et al. 1999. Effects of long-term treatment of colon adenocarcinoma with crocin, a carotenoid

from saffron (*Crocus sativus*): An experimental study in the rat. Nutrition and Cancer 35(2): 120-126.

Ghazi H, Hirotsugu M, Norio N, Masao H, Nobuko K, et al. 2000. Inhibitory effects of Sudanese medicinal plant extracts on hepatitis C virus (HCV) protease. Phytotherapy Research 14(7): 510-516.

Ghelardini C, Galeotti N, Mannelli LDC, Mazzanti G, Bartolini A. 2001. Local anaesthetic activity of beta-caryophyllene. Il Farmaco 56(5-7): 387-389.

Giri J, Devi TKS, Meerarani S. 1984. Effect of ginger on serum cholesterol levels. Indian Journal of Nutrition and Dietetics 21(12): 433-436.

Government of India. 1989. Turmeric. In: Ayurvedic pharmacopoeia of India. Department of Health, Government of India, New Delhi. p. 45-46.

Graf J. 2000. Herbal anti-inflammatory agents for skin disease. Skin Therapy Letter 5: 3-5.

Guillen MD, Manzanos MJ. 1996. A study of several parts of the plant *Foeniculum vulgare* as a source of compounds with industrial interest. Food Research International 29(1): 85-88.

Hammer KA, Carson CF, Riley TV. 1999. Antimicrobial activity of essential oils and other plant extracts. Journal of Applied Microbiology 86(6): 985-990.

Hastings CW, Barnes DJ, Daley CA. 2001. Dietary supplement for nutritionally promoting healthy joint function. U.S. Patent 6224871 B1 20010501. p. 4.

Hinneburg I, Damien HJ, Hiltunen R. 2006. Antioxidant activities of extracts from selected culinary herbs and spices. Food Chemistry 97(1): 122-129.

Ho SH, Ma Y, Huang Y, Hook F. 1997. Anethole, a potential insecticide from *Illicium verum* against two stored product insects. International Pest Control 39(2): 50-51.

Hosseinzadeh H, Karimi G, Niapour M. 2004. Antidepressant effects of *Crocus sativus* stigma extracts and its constituents, crocin and safranal, in mice. Acta Horticulturae 650: 435.

Hosseinzadeh H, Khosravan V. 2002. Anticonvulsant effects of aqueous and ethanolic extracts of Crocus sativus L. stigmas in mice. Archives of Internal Medicine 5: 44.

Hosseinzadeh H, Younesi H. 2002. Antinociceptive and anti-inflammatory effects of *Crocus sativus* L. stigma and petal extracts in mice. BMC Pharmacology 2: 7.

Hsieh PC, Mau JL, Huang SH. 2001. Antimicrobial effect of various combinations of plant extracts. Food Microbiology 18(1): 35-43.

Hussein G, Miyashiro H, Nakamura N, Hattori M, Kakiuchi N, et al. 2000. Inhibitory effects of Sudanese medicinal plant extracts on hepatitis C virus (HCV) protease. Phytotherapy Research 14(7): 510-516.

Ilori M, Sheteolu AO, Omonigbehin EA, Adeneye AA. 1996. Antidiarrhoeal activities of *Ocimum gratissimum* (Lamiaceae). Journal of Diarrhoeal Diseases Research 14(4): 283-285.

Inahata K, Nakasugi T, Matsumura S, Nakahara T. 1996. Food for preventing dental diseases. Japanese Kokai Tokkyo Koho. p. 6.

Inahata K, Nakasugi T, Matsumura S, Nakahara, S. 1997. Foods containing *Murraya* extracts for prevention and control of bad breath. Japanese Kokai Tokkyo Koho. p. 8.

Inouye S, Takizawa T, Yamaguchi H. 2001. Antibacterial activity of essential oils and their major constituents against respiratory tract pathogens by gaseous contact. Journal of Antimicrobial Chemotherapy 47(5): 565-573.

Inouye S, Tsuruoka T, Watanabe M, Takeo K, Akao M, et al. 2000. Inhibitory effect of essential oils on apical growth of *Aspergillus fumigatus* by vapour contact. Mycoses 43(1-2): 17-23.

Isao K, Masaki H, Hisae M. 1991. Antimicrobial activity of flavor components of cardamom *Elettaria cardamomum* (Zingiberaceae) seed. Journal of Agricultural and Food Chemistry 39(11): 1984-1986.

Jafri, MA, Farah, Javed K, Singh S. 2001. Evaluation of the gastric antiulcerogenic effect of large cardamom (fruits of *Amomum subulatum* Roxb). Journal of Ethnopharmacology 75(2-3): 89-94.

Jampani H, Ellis T, Newman JL. 2001. Therapeutic antimicrobial compositions. Patent Cooperation Treaty International Application. p. 58.

Jannu LN, Hussain SP, Rao AR. 1991. Chemopreventive action of mace (*Myristica fragrans,* Houtt) on DMBA-induced papillomagenesis in the skin of mice. Cancer Letters 56(1): 59-63.

Japanese Ministry of Health and Welfare. 1993. The Japanese standards for herbal medicines. Yakuji Nippo, Tokyo. p. 279.

Jessie SW, Krishnakantha TP. 2005. Inhibition of human platelet aggregation and membrane lipid peroxidation by food spice, saffron. Molecular and Cellular Biochemistry 278(1-2): 59-63.

Joseph S, Peter KV. 1985. Curry leaf *Murraya koenigii:* A perennial, nutritious, leafy vegetable. Economic Botany 39(1): 68-73.

Jung GD, Yang JY, Song ES, Par JW. 2001. Stimulation of melanogenesis by glycyrrhizin in B16 melanoma cells. Experimental and Molecular Medicine 33(3): 131-135.

Kamei T, Kumano H, Iwata K, Nariai Y, Matsumoto T. 2000. The effect of a traditional Chinese prescription for a case of lung carcinoma. Journal of Alternative and Complementary Medicine 6(6): 557-559.

Karthikeyan K, Ravichandran P, Govindasamy S. 1999. Chemopreventive effect of *Ocimum sanctum* on DMBA-induced hamster buccal pouch carcinogenesis. Oral Oncology 35(1): 112-119.

Kazuho A, Hiroshi S. 2000. Effects of saffron extract and its constituent crocin on learning behavior and long-term potentiation. Phytotherapy Research 14(3): 149-152.

Kelloff GJ, Crowell JA, Steele VE, Lubet RA, Malone WA, et al. 2000. Progress in cancer chemoprevention: Development of diet-derived chemopreventive agents. Journal of Nutrition 130(2 suppl.): 467S-471S.

Khan BA, Abraham A, Leelamma S. 1995. Hypoglycemic action of *Murraya koenigii* (curry leaf) and *Brassica juncea* (mustard): Mechanism of action. Indian Journal of Biochemistry and Biophysics 32: 106-108.

Khan BA, Abraham A, Leelamma S. 1996a. Hypolipidic properties: Biochemical response in rats to the addition of curry leaf (*Murraya koenigii*) and mustard seeds (*Brassica juncea*) to the diet. Plant Foods for Human Nutrition 49(4): 295-299.

Khan BA, Abraham A, Leelamma S. 1996b. Reduction in the mean number of tumors in colon carcinogenesis. *Murraya koenigii* and *Brassica juncea*-alterations on lipid profile in 1-2 dimethyl hydrazine induced colon carcinogenesis. Investigational New Drugs 14: 365-369.

Khan BA, Abraham A, Leelamma S. 1997. Anti-oxidant effects of curry leaf, *Murraya koenigii* and mustard seeds, *Brassica juncea* in rats fed with high fat diet. Indian Journal of Experimental Biology 35(2): 148-150.

Khayyal MT, El-Ghazaly MA, Kenawy SA, Seif-El-Nasr M, Mahran LG, et al. 2001. Antiulcerogenic effect of some gastrointestinally acting plant extracts and their combination. Arzeneimittel-Forschung 51(7): 545-553.

Khushbaktova ZA, Syrov VN, Batirov EK. 1991. Effects of flavonoids on the course of hyperlipidemia and atherosclerosis. Khimiko-Farmatsevticheskii Zhurnal 25(4): 53-57.

Konovalova GG, Tikhaze AK, Lankin VZ. 2000. Antioxidant activity of parapharmaceutics containing natural inhibitors of free radical processes. Bulletin of Experimental Biology and Medicine 130(7): 658-660.

Lachowicz KJ, Jones GP, Briggs DR, Bienvenu FE, Wan J, et al. 1998. The synergistic preservative effects of the essential oils of sweet basil (*Ocimum basilicum* L.) against acid-tolerant food microflora. Letters in Applied Microbiology 26(3): 209-214.

Lee CK, Moon KH, Kim H. 1999. Antibiotics resistance inhibition by natural products. Proceedings of 3rd International Symposium: Recent Advances in Natural Products Research. Seoul National University Press, Seopul, Republic of Korea. p. 74-92.

Lee KG, Shibamoto T. 2001. Antioxidant property of aroma extract isolated from clove buds [*Syzygium aromaticum* (L.) Merr. et Perry]. Food Chemistry 74(4): 443-448.

Luper S. 1999. A review of plants used in the treatment of liver disease: Part two. Alternative Medicine Review 4(3): 178-188.

Malini T, Vanithakumari G, Megala N, Anusya S, Devi K, et al. 1985. Effect of *Foeniculum vulgare* Mill. seed extract on the genital organs of male and female rats. Indian Journal of Physiology and Pharmacology 29(1): 21-26.

Martini H, Weidenborner M, Adams S, Kunz B. 1996. Eugenol and carvacrol: The main fungicidal compounds in clove. Italian Journal of Food Science 1: 63-67.

Matsumura S, Inahata K, Nakasugi T, Yoshikawa A. 1994. Microbicidal air fresheners for bathroom. Japanese Kokai Tokkyo Koho. p. 5.

Maulik G, Maulik N, Bhandari V, Kagan VE, Pakrashi S, et al. 1997. Evaluation of antioxidant effectiveness of a few herbal plants. Free Radical Research 27(2): 221-228.

Milic BL, Milic NB. 1998. Protective effects of spice plants on mutagenesis. Phytotherapy Research 12(Suppl. 1): S3-S6.

Miyazawa M, Ota H, Ishikawa Y, Kameoka H. 1993. An insecticidal compound from *Illicium verum*. Chemistry Express 8(7): 437-440.

Murakami H, Ohigashi, Koshimizu K. 1994. Possible anti-tumour promoting properties of traditional Thai food items and some of their active constituents. Asia Pacific Journal of Clinical Nutrition 3: 185-191.

Na MK, An RB, Lee SM, Hong ND, Yoo JK, et al. 2001. Screening of crude drugs for antioxidative activity. Saengyak Hakhoechi 32(2): 108-115.
Nair SC, Pannikar B, Panikkar KR. 1991. Antitumour activity of saffron (*Crocus sativus*). Cancer Letters 57: 109.
Nakajima I. 1989. *Myristica fragrans* extract as neoplasm inhibitor. Japanese Kokai Tokkyo Koho. p. 5.
Nakamura T, Okuyama E, Yamazaki M. 1996. Neurotropic components from star anise (*Illicium verum* Hook. fil.). Chemical and Pharmaceutical Bulletin 44(10): 1908-1914.
Nakano Y, Matsunaga H, Saita T, Mori M, Katano M, et al. 1998. Antiproliferative constituents in Umbelliferae plants II. Screening for polyacetylenes in some Umbelliferae plants, and isolation of panaxynol and falcarindiol from the root of *Heracleum moellendorffii*. Biological and Pharmaceutical Bulletin 21(3): 257-261.
Nakatani N, Inatani R, Ohta H, Nishioka A. 1986. Chemical constituents of peppers (*Piper* spp.) and application to food preservation: Naturally occurring antioxidative compounds. Environmental Health Perspectives 67: 135-142.
Nakayama R, Kikuzaki H, Nakatani N, Horiuchi H. 1996. Antioxidative activity of constituents from fennel seeds. Nippon Kasei Gakkaishi 47(12): 1193-1199.
Nalini N, Sabitha K, Viswanathan P, Menon VP. 1998. Influence of spices on the bacterial (enzyme) activity in experimental colon cancer. Journal of Ethnopharmacology 62(1): 15-24.
Namba T, Tsunezuka M, Dissanayake DMRB, Pilapitiya U, Saito K, et al. 1985. Studies on dental caries prevention by traditional medicines (Part VII). Screening of ayurvedic medicines for anti-plaque action. Shoyakugaku Zasshi 39(2): 146-153.
Nutan MTH, Hasnat A, Rashid, MA. 1998. Antibacterial and cytotoxic activities of *Murraya koenigii*. Fitoterapia 69(2): 173-175.
Olajide OA, Ajayi FF, Ekhelar AI, Awe SO, Makinde JM, Alada AR. 1999. Biological effects of *Myristica fragrans* (nutmeg) extract. Phytotherapy Research 13(4): 344-345.
Onawunmi GO, Yisak WA, Ogunlana EO, Viana GS, Vale TG, et al. 1984. Antibacterial constituents in the essential oil of *Cymbopogon citrates* (DC.) Stapf. Journal of Ethnopharmacology 12(3): 279-286.
Onderoglu S, Sozer S, Erbil KM, Ortac R, Lermioglu F. 1999. The evaluation of long-term effects of cinnamon bark and olive leaf on toxicity induced by streptozotocin administration to rats. Journal of Pharmacy and Pharmacology 51(11): 1305-1312.
Oshima K, Mitsunaga T. 1999. Glucosyl transferase inhibitors from plant extracts for dentifrice. Japanese Kokai Tokkyo Koho. p. 9.
Ozaki Y, Soedigdo S, Wattimena YR, Suganda AG. 1989. Antiinflammatory effect of mace, aril of *Myristica fragrans* Houtt., and its active principles. Japanese Journal of Pharmacology 49(2): 155-163.
Palaniswamy UR, Caporuscio C, Stuart JD. 2003. A chemical analysis of antioxidant vitamins in fresh curry leaf (*Murraya koenigii*) by reversed phase HPLC with UV detection. Acta Horticulturae 620: 475-478.

Piccaglia R, Marotti M, Giovanelli E, Deans SG, Eaglesham E. 1993. Antibacterial and antioxidant properties of Mediterranean aromatic plants. Industrial Crops and Products 2(1): 47-50.

Platel K, Srinivasan K. 1996. Influence of dietary spices or their active principles on digestive enzymes of small intestinal mucosa in rats. International Journal of Food Science and Nutrition 47(1): 55-59.

Platel K, Srinivasan K. 2000. Influence of dietary spices and their active principles on pancreatic digestive enzymes in albino rats. Nahrung 44(1): 42-46.

Platel K, Srinivasan K. 2001. A study of the digestive stimulant action of select spices in experimental rats. Journal of Food Science and Technology 38(4): 358-361.

Polasa K, Raghuram TC, Krishna TP, Krishnaswamy K. 1992. Effect of turmeric on urinary mutagens in smokers. Mutagenesis 7: 107-109.

Ponglux D, Wongseripipatana S, Phadungcharoen T, Rungrungsri N, Likhitwitayawuid K. 1987. Medicinal plants. Victory Power Point Corp, Bangkok.

Pradeep UK, Geervani P, Eggum BO. 1993. Common Indian spices: Nutrient composition, consumption and contribution to dietary value. Plant Foods for Human Nutrition 44(2): 137-148.

Pradhan KJ, Variyar PS, Bandekar JR. 1999. Antimicrobial activity of novel phenolic compounds from green pepper (*Piper nigrum* L.). Lebensmittel-Wissenschaft und -Technologie 32(2): 121-123.

Premkumar K, Thirunavukkarasu C, Abraham SK, Santhiya ST, Ramesh A. 2006. Protective effect of saffron (*Crocus sativus* L.) aqueous extract against genetic damage induced by anti-tumor agents in mice. Human and Experimental Toxicology 25(2): 79-84.

Puatanachokchai R, Kishida H, Denda A, Murata N, Konishi Y, et al. 2002. Inhibitory effects of lemon grass (*Cymbopogon citratus*, Stapf) extract on the early phase of hepatocarcinogenesis after initiation with diethylnitrosamine in male Fischer 344 rats. Cancer Letters 183(1): 9-15.

Rahman AU, Choudhary MI, Farooq AAA, Iqbal MZ, Demirci B, et al. 2000. Antifungal activities and essential oil constituents of some spices from Pakistan. Journal of the Chemical Society of Pakistan 22(1): 60-65.

Rai V, Iyer U, Mani UV. 1997. Effect of Tulasi (*Ocimum sanctum*) leaf powder supplementation on blood sugar levels, serum lipids and tissue lipids in diabetic rats. Plant Foods for Human Nutrition 50(1): 9-16.

Ramadan A, Afifi NA, Fathy MM, El-Kashoury EA, El-Naeneey EV. 1994. Some pharmacodynamic effects and antimicrobial activity of essential oils of certain plants used in Egyptian folk medicine. Veterinary Medical Journal Giza 42: 263-270.

Rao L, Jagan M, Ramalakshmi K, Borse BB, Raghavan B. 2007. Antioxidant and radical-scavenging carbazole alkaloids from the oleoresin of curry leaf (*Murraya koenigii* Spreng.). Food Chemistry 100(2): 742-747.

Ravindra JK, Srivastava SD. 1992. Anti-inflammatory, anti-microbial and phytochemical studies on the stem bark of *Murraya koenigii*. Proceedings of the Indian National Academy of Science A 62(1): 5-9.

Razina TG, Zueva EP, Amosova EN, Krylova SG. 2000. Medicinal herb preparations as a means of additional therapy: Experimental oncology evidence. Eksperimental'naya i Klinicheskaya Farmakologiya 63(5): 59-61.

Reddy PRK, Reddy SJ. 1996. Assessment of macronutrients in some selected leafy samples. Ecology, Environment and Conservation 2: 41-44.

Ríos JL, Recio MC, Giner RM, Máez S. 1998. An update review of saffron and its active constituents. Phytotherapy Research 10: 189.

Roth GN, Chandra A, Nair MG. 1998. Novel bioactivities of *Curcuma longa* constituents. Journal of Natural Products 61: 542-545.

Ruberto G, Baratta MT, Deans SG, Dorman HJD. 2000. Antioxidant and antimicrobial activity of *Foeniculum vulgare* and *Crithmum maritimum* essential oils. Planta Medica 66(8): 687-693.

Saleem M, Alam A, Sultana S. 2001. Asafoetida inhibits early events of carcinogenesis: A chemopreventive study. Life Sciences 68(16): 1913-1921.

Salomi MJ, Nair SC, Panikkar KR. 1991. Inhibitory effects of *Nigella sativa* and saffron (*Crocus sativus*) on chemical carcinogenesis in mice. Nutrition and Cancer 16(1): 67-72.

Selvam R, Subramanian L, Gayathri R, Angayarkanni N. 1995. The anti-oxidant activity of turmeric (*Curcuma longa*). Journal of Ethnopharmacology 47: 59-67.

Shang S, Wu X, Yang L, Lan S. 1994. Screening antioxidants from Chinese herb medicines. Tianran Chanwu Yanjiu Yu Kaifa 6(1): 36-39.

Shankel DM, Clarke CH. 1990. Specificity of antimutagens against chemical mutagens in microbial systems. Basic Life Sciences 52: 457-460.

Sharma A, Mathur R, Dixit VP. 1995. Prevention of hypercholesterolemia and atherosclerosis in rabbits after supplementation of *Myristica fragrans* seed extract. Indian Journal of Physiology and Pharmacology 39(4): 407-410.

Shcherbanovsky LR, Kapelev IG. 1975. Volatile oil of *Anethum graveolens* L. as an inhibitor of yeast and lactic acid bacteria. Prikladnaya Biokhimiya i Mikrobiologiya 11(3): 476-477.

Shobana S, Naidu KA. 2000. Antioxidant activity of selected Indian spices. Prostaglandins, Leukotrienes, and Essential Fatty Acids 62: 107-110.

Singh G, Marimuthu P, Murali HS, Bawa AS. 2005. Antioxidative and antibacterial potentials of essential oils and extracts isolated from various spice materials. Journal of Food Safety 25(2): 130-145.

Singh HB, Srivastava M, Singh AB, Srivastava AK. 1995. Cinnamon bark oil, a potent fungitoxicant against fungi causing respiratory tract mycoses. Allergy 50(12): 995-999.

Singh S. 1999. Evaluation of gastric anti-ulcer activity of fixed oil of *Ocimum basilicum* Linn. and its possible mechanism of action. Indian Journal of Experimental Biology 37(3): 253-257.

Singh S, Majumdar DK. 1999. Evaluation of the gastric antiulcer activity of fixed oil of *Ocimum sanctum* (Holy Basil). Journal of Ethnopharmacology 65(1): 13-19.

Singhal PC, Joshi LD. 1983. Glycemic and cholesterolemic role of ginger and till. Journal of Scientific Research in Plants and Medicines 4(3): 32-34.

Soma R, Ikeda M, Morise T, Miyamori I, Takeda R. 1994. Effect of glycyrrhizin on cortisol metabolism in humans. Endocrine Regulations 28: 31-34.

Soni KB, Lahiri M, Chackradeo P, Bhide SV, Kuttan R. 1997. Protective effect of food additives on aflatoxin-induced mutagenicity and hepatocarcinogenicity. Cancer Letters 115(2): 129-133.

Soni KB, Rajan A, Kuttan R. 1992. Reversal of aflatoxin induced liver damage by turmeric and curcumin. Cancer Letters 66(2): 115-121.

Soudamini KK, Unnikrishnan MC, Sukumaran K, Kuttan R. 1995. Mutagenicity and anti-mutagenicity of selected spices. Indian Journal of Physiology and Pharmacology 39(4): 347-353.

Srinivasan MR, Srinivasan K. 1995. Hypocholesterolemic efficacy of garlic-smelling flower *Adenocalymma alliaceum* Miers. in experimental rats. Indian Journal of Experimental Biology 33(1): 64-66.

Sugiura M. 1999. Pharmacology of *Crocus sativus* on the central nervous system as a health food for dementia. Food Style 3(6): 63-68.

Sujatha R, Srinivas L. 1995. Modulation of lipid peroxidation by dietary components. Toxicology in Vitro 9(3): 231-236.

Sung CK, Kimura T, But PPH, Guo JX. 1997. Turmeric. In: Sung CK, Kimura T, But PPH, Guo JX, editors, International collation of traditional and folk medicine. World Scientific, Singapore. p. 207-208.

Surh Y. 1999. Molecular mechanisms of chemopreventive effects of selected dietary and medicinal phenolic substances. Mutation Research 428(1-2): 305-327.

Tanaka R, Morishima M, Shibuya K, Maeda N, Shimoda S, et al. 1998. The effect of Murraya (*Murraya koenigii*) extract on experimental caries. Shika Yakubutsu Ryoho 17(3): 161-166.

Terasawa N, Yamazaki N, Fukui, Y. 2001. Antioxidative activity of water extracts of herbs. Nippon Shokuhin Kagaku Kogaku Kaishi 48(2): 99-104.

Thyagaraja N, Hosono A. 1996. Effect of spice extract on fungal inhibition. Journal of Food Science and Technology 29(3): 286-288.

Tkachenko KG, Kazarinova NV, Muzychenko LM, Shurgaya AM, Pavlova OV, et al. 1999. Antibiotic properties of essential oils of some plant species. Rastitel'nye Resursy 35(3): 11-24.

Tu G, Fang Q, Yuan S. 1992. Pharmacopoeia of the People's Republic of China (English edition). Guangdong Science and Technology Press, Beijing. p. 202-203.

Unnikrishnan MC, Kuttan R. 1990. Tumour reducing and anticarcinogenic activity of selected spices. Cancer Letters 51(1): 85-89.

Van Dau N, Ham NN, Khac DH, Lam N, Son PT, et al. 1998. The effects of traditional drug, turmeric (*Curcuma longa*), and placebo on the healing of duodenal ulcer. Phytomedicine 5: 29-34.

Van GC, Cox PA. 1994. Ethnobotany of nutmeg in the Spice Islands. Journal of Ethnopharmacology 42(2): 117-124.

Vats V, Grover JK, Rathi SS. 2002. Evaluation of anti-hyperglycemic and hypoglycemic effect of *Trigonella foenum-graecum* Linn, *Ocimum sanctum* Linn and *Pterocarpus marsupium* Linn in normal and alloxanized diabetic rats. Journal of Ethnopharmacology 79(1): 95-100.

Vaya J, Belinky PA, Aviram M. 1997. Antioxidant constituents from licorice roots: Isolation, structure elucidation and antioxidative capacity toward LDL oxidation. Free Radical Biology and Medicine 23(2): 302-313.

Verma SK, Bordia A. 1998. Antioxidant property of saffron in man. Indian Journal of Medical Sciences 52(5): 205-207.
Viana GSB, Vale TG, Pinho RSN, Matos FJA. 2000. Antinociceptive effect of the essential oil from *Cymbopogon citrates* in mice. Journal of Ethnopharmacology 70(3): 323-327.
Vijayakumar RS, Surya D, Nalini N. 2004. Antioxidant efficacy of black pepper (*Piper nigrum* L.) and piperine in rats with high fat diet induced oxidative stress. Redox Report 9(2): 105-110.
Vijayakumar RS, Surya D, Senthilkumar R, Nalini N. 2002. Hypolipidemic effect of black pepper (*Piper nigrum* Linn.) in rats fed high fat diet. Journal of Clinical Biochemistry and Nutrition 32: 31-42.
Vinitketkumnuen U, Puatanachokchai R, Lertprasertsuke N, Kongtawelert P, Picha P, et al. 1996. Antimutagenicity and anti-tumor activity of lemon grass. Mutation Research/Environmental Mutagenesis and Related Subjects 359(3): 200-201.
Wang ZY. 1994. Anticarcinogenesis of licorice and its major triterpenoid constituents. In: Huang MT, Osawa T, Hou CT, Roson RT, editors. Food phytochemicals for cancer prevention II teas, spices, and herbs. ACS Symp Ser 547. Oxford University Press, Oxford, UK. p. 329-334.
Whorwood CB, Shepard MC, Stewart PM. 1993. Licorice inhibits 11-hydroxysteroid dehydrogenase messenger ribonucleic acid levels and potentiates glucocorticoid hormone action. Endocrinology 132: 2287-2292.
Williams GO, Omoh LE. 1996. Mitotic effects of the aqueous leaf extract of *Cymbopogon citrates* in *Allium cepa* root tips. Cytobios 87(350): 161-168.
Winterhalter P, Straubinger M. 2000. Saffron-renewed interest in an ancient spice. Food Reviews International 16(1): 39-59.
Xie JT, Chang WT, Wang CZ, Mehendale SR, Li J, et al. 2006. Curry leaf (*Murraya koenigii* Spreng) reduces blood cholesterol and glucose levels in ob/ob mice. American Journal of Clinical Medicine 34(2): 279-284.
Xuan B, Zhou YH, Li N, Min ZD, Chiou GCY. 1999. Effects of crocin analogs on ocular blood flow and retinal function. Journal of Ocular Pharmacology and Therapeutics 15(2): 143-152.
Yamasaki K, Nakano M, Kawahata T, Mori H, Otake T, et al. 1998. Anti-HIV-1 activity of herbs in Labiatae. Biological and Pharmaceutical Bulletin 21(8): 829-833.
Yamashiki M, Nishimura A, Huang XX, Nobori T, Sakaguchi S, et al. 1999. Effects of the Japanese herbalmedicine "Sho-saiko-to" (TJ-9) on interleukin-12 production in patients with HCV-positive liver cirrhosis. Developmental Immunology 7(1): 17-22.
Yazdanparast R, Alavi M. 2001. Antihyperlipidaemic and antihypercholesterolaemic effects of *Anethum graveolens* leaves after the removal of furocoumarins. Cytobios 105(410): 185-191.
Zhang T, Li Q, Zhang J. 1999. Cigarette with traditional Chinese medicines. Faming Zhuanli Shenqing Gongkai Shuomingshu. p. 6.
Zheng GQ, Kenney PM, Lam LK. 1992. Anethofuran, carvone, and limonene: Potential cancer chemopreventive agents from dill weed oil and caraway oil. Planta Medica 58(4): 338-341.

Chapter 7

Herbal Beverages and Subsequents (Postprandial Treats)

In all cultures people frequently enjoy postprandial treats, which are termed "subsequents" here, that follow dessert or the end of a meal; in the West, maybe drinking a cup of coffee or tea or smoking a cigarette, or in the East, chewing a few seeds of fennel, betel nuts, or *pan,* which is a mix of betel leaf, betel nut, and a touch of slaked lime (calcium hydroxide) and often flavored with other spices and additives. These treats, variously called "digestives," "afters," or simply refreshments, are termed subsequents here to denote that it is something taken soon after a meal for whatever reason(s) given by humans, ranging from satisfactory completion of a meal to stimulate digestion or to refresh breath. Subsequents are believed in some cultures to have a beneficial effect when used in appropriate quantities: benefits include stimulation, a sense of well-being, enhanced sexual well-being, and aphrodisiac effects. Along with the more conventional food items covered in previous chapters, such items, especially those used as part of a culture's diet, need to be studied and their effects on humans examined in detail. For example, beverages such as tea are known for their antioxidant properties and polyphenolic content, which prevents degenerative diseases such as cancer and heart disease. As with any plant species, these subsequents also contain a complex array of phytochemicals, and their possible effect on health promotion and disease prevention is an important consideration for a dietitian.

Asian Crops and Human Dietetics

doi:10.1300/5504_07

TEA

Tea (*Camellia sinensis;* Theaceae) is indigenous to Asia, specifically China. It dates back to 2737 BC, when the Chinese emperor Chen-Nung discovered by chance that an infusion of wild tea bush leaves was remarkably reviving. Since the fourth century tea has been an integral part of life in China and a major feature of important ceremonies. It spread to Japan in early AD 1400, and the tea ceremony, a ritual central to Japanese culture, came into being. Tea was introduced to Europe when the French East India Company brought tea to Holland in 1610, France in 1636, and England in 1650. Tea drinking has become part of English daily social custom. In the United States, tea was very popular with the early colonials and a valuable source of income to the British government. In 1773 as an act of protest against high taxes on tea, rebels dumped tea cargo in the ocean during the famous Boston tea party. Tea continues to be a beverage choice for many, especially after learning of the health benefits documented in many epidemiological and clinical studies.

Major tea growers of the world include India, China, Sri Lanka, Korea, parts of Africa, and South America. Tea is an evergreen tree that can grow to more than 10 m; in cultivation it is pruned frequently to maintain a more manageable height of approximately 1.5 m. Tea plants bear dark glossy green leaves and fragrant white flowers. The tender buds (pekoe) are harvested with two terminal leaves and processed to obtain the commercial tea. Three major categories of tea are derived from *C. sinensis* by different processing methods: black teas (fermented), oolong teas (semifermented), and green teas (unfermented). The longer the fermentation, the more caffeine is contained in the final product. Black tea is popular in India and manufactured by two methods: orthodox and crush-tear-curl (CTC). The orthodox method involves rolling the tea leaves in a machine to break them and release the phytochemicals; the CTC method uses a special machine to crush and break the leaves.

In traditional Indian and Chinese medicine, green tea is used as a stimulant, diuretic, and astringent, and to regulate body temperature, promote digestion, improve mental function, and cure headaches. In Chinese medicine, tea is used to exert an influence over the nervous system, giving a feeling of comfort and exhilaration. Tea is used internally to treat diarrhea, hepatitis, and gastroenteritis and externally to treat cuts, burns, bruises, insect bites, and swellings. Because of the caffeine tea contains, excessive tea drinking can lead to dizziness, constipation, indigestion, palpitations, and insomnia.

The active constituents in tea that promote good health are polyphenols, particularly the catechin called epigallocatechin gallate (EGCG; Graham 1992). A typical cup of tea prepared in a proportion of 1 g leaf to 100 mL water in a three-minute brew contains 30% to 42% catechins and 3% to 6% caffeine (Balentine et al. 1997). Herbal teas are made of leaves of plant species other than *C. sinensis* and may not contain any catechins. Tablets and capsules containing extracts of EGCG are available, some of which provide up to 97% polyphenol content, or the equivalent of drinking four cups of tea. Research demonstrates that green tea protects against cardiovascular disease in many ways: it lowers total cholesterol levels and improves the cholesterol profile (the ratio of low-density to high-density lipoprotein), reduces platelet aggregation, and lowers blood pressure (Kono et al. 1992; Sagesaka-Mitane et al. 1990; Stensvold et al. 1992; Yamaguchi et al. 1991). The polyphenols in green tea have been shown to be effective antioxidants (Serafini et al. 1996), lessen the risk of cancers of several sites, stimulate the production of several immune system cells, and have antibacterial properties, specifically, against the bacteria that cause dental plaque (Hamilton-Miller 1995, 2001; Stoner and Mukhtar 1995; You 1993). Tea drinking can also offer protection against liver damage (Imai and Nakachi 1995) and osteoporosis (Hegarty et al. 2000).

ARECA NUT

The areca nut (*Areca catechu;* Palmae, or Arecaceae) is native to Indonesian and southeastern Asian islands. Also known as betel nut, the areca nut is popular as an ingredient in the betel leaf masticatory preparation called *pan* in Hindi. *Pan* is called *tambula* in Sanskrit, the term derived from *tamra,* "copper," the color the mouth is stained after chewing *pan*.

Areca nut is a cash crop in many Asian countries and cultivated widely in India, Pakistan, Indonesia, Sri Lanka, Africa, Egypt, Turkey, and the Pacific Islands. It is also grown as ornamental landscape and specimen trees. The areca palm grows 9 to 18 m (30-60 ft). The trunk of the tree is unbranched, slender, and erect with conspicuous gray rings. The leaves are dark green, pinnate, and up to 1.8 to 2.4 m (6-8 ft) long, with the leaflets irregularly serrate at the tip. The flowering stems grow from the trunk well below the leaves. The flowers are borne on a branching spadix, with the female flowers occurring solitarily at the base and the small, white, fragrant male flowers surrounding the solitary female flowers. The fruits are ovoid berries, green in the immature state and turning orange as they ripen, containing brownish, conical seeds.

According to Sushrata, the ancient Indian medicine man, *pan* keeps the mouth clean and prevents bacterial infections, aids in digestion, and purifies blood. Chewing *pan* is an age-old tradition deeply rooted in India, integrated within social and religious systems, and used as a magical charm to protect against the evil eye and ward off demons. Betel leaf and areca nut are customary and invaluable offerings to guests, to gods during worship, and to signify auspicious occasions in Hindu ceremonies and religious occasions.

In India, Pakistan, Malaysia, Sri Lanka, and Bangladesh, it is considered a subsequent or a postprandial digestive and is chewed either alone, wrapped with slaked lime in betel leaf, or with tobacco. Depending on individual and local preferences, condiments, sweetening agents, and spices are also added to the preparations. The areca nut is used in Ayurvedic medicine as an aphrodisiac and a nerve stimulator to enhance the feeling of well-being, and it is prescribed for stomach and urinary problems. Areca nut is used medicinally for intestinal worms and dysentery.

Scientific reports on the effects of areca nut consumption on human health and well-being vary, with some on its medicinal and curative properties and several on its harmful and toxicological effects, depending on frequency and quantity of consumption. Areca nut chewing has long been a tradition in India, Taiwan, and many other parts of southern Asia. Areca nut contains the alkaloids arecaine, arecoline, arecaidine, and guvacine. Consumption of areca nuts is traditionally known to produce mental alertness and dizziness, and it is sometimes used as a euphoric agent. Long-term consumption of areca nuts can cause hypertension, palpitation, sweating, vomiting, and abdominal pain (Huang et al. 1997).

Evidence is growing that areca nut chewing is a risk factor for malabsorption of nutrients (Kumar et al. 2000) and development of diabetes (Mannan et al. 2000), and it may cause oral submucosal fibrosis, a premalignant condition associated with oral cancer (Noor 1987). Areca nut chewing is implicated as a risk factor for cardiovascular diseases and asthma (Sharan and Wary 1992; Taylor et al. 1992; Wary and Sharan 1991). There are numerous case reports of oral cancer developing among individuals who use areca nut but not tobacco in any form, and it has been associated with oral leukoplakia and squamous cell carcinoma (Norton 1998).

Some studies demonstrate no obvious increase in oral cancer risk among areca nut chewers compared with those who do not chew the nut (Stich et al. 1986). Other research has reported that the beneficial and medicinal properties of the areca nut include that it is effective as an antihypertensive (Inokuchi et al. 1986), an antinociceptive (Ghelardini et al. 2001), and an antidepressant agent (Dar and Khatoon 2000).

BETEL LEAF

The betel leaf (*Piper betle;* Piperaceae) is native to southern Asia, where it has been chewed since antiquity. Use of betel leaves can be traced as far back as 2,000 years. Betel leaves are commonly found in southern Asian and Indian homes and are chewed almost daily by young and old alike.

The betel plant is a vine climbing to 4.5 m (15 ft), with rounded woody stems. The leaves are 10 to 15 cm long and 5 to 7.5 cm wide, ovate or heart shaped at the base with an acute tip. The leaves are green, smooth, glabrous, and somewhat leathery, with a prominent midrib and veins. The long-petioled leaves are arranged alternately on the stem. The leaves have a pungent odor and taste. The inflorescence is borne on leaf axils. The fruits are 10 to 15 cm long, embedded in rachis, and joined.

Traditionally, slices of areca nuts (betel palm seeds) spread on a betel leaf, smeared with slaked lime, optionally filled with other aromatic flavorings, and rolled up as *pan,* are very common in India. Betel leaves are offered to guests as a sign of welcome and mark the auspicious nature of a visit and occasion; betel chewing after a meal is customary for some people as a sign of satisfaction of having eaten a good meal and is believed to aid digestion. Betel chewing is the fourth most common addiction, after nicotine, alcohol, and caffeine, which are used more widely than betel.

Betel vine is extensively grown in Southeast Asia, Madagascar, and east Africa for its leaves. In the Indian Subcontinent, the leaves are rolled as *pan* and chewed habitually as a digestive and nerve stimulant. The root, leaves, and seeds are all used medicinally. The leaves and stalks are aromatic and considered to be a tonic, an expectorant, and a stimulant. The leaf extracts are considered antibacterial and are used externally to clean cuts and to treat wounds and skin ulcers. The betel leaf is reported to be a good source of minerals (Calcium, magnesium, and phosphorus) and vitamins (vitamin C, beta-carotene, thiamine, riboflavin, and niacin) (Mehrotra et al. 1999).

Betel leaves have been used since ancient times in Ayurvedic medicine as an aphrodisiac and in home remedies for cough, the common cold, and congestion of the throat and chest. Betel leaf juice has diuretic, analgesic, and cooling properties, and it is applied to the temples and forehead to relieve fever and headache. Betel leaves soaked in mustard oil, coconut oil, or sesame oil and warmed and applied to the chest relieve cough and ease difficulty in breathing. Betel leaves are beneficial in the treatment of inflammation such as arthritis; warmed and smeared with castor oil they are used externally to heal inflammation. Betel leaves can be used to heal wounds and boils. The stalk of a betel leaf dipped in castor oil is used as a homemade suppository to relieve constipation. The mild stimulation and intoxication experienced after chewing pan led to its use in some religious ceremonies and wedding rituals for welcoming the husband and as a symbol of prosperity.

The active principle responsible for the feeling of intoxication is the alkaloid arecoline in the areca nut. Scientists report that the chewing of betel nut may relieve many of the symptoms of schizophrenia. Cancer of the mouth and lips has been found to be more frequent in areas where betel chewing is a widely prevalent habit. Other conditions reported after excessive pan chewing included dyspepsia, pyorrhea, and cancers of the tongue and cheeks (Jeng et al. 1999; Norton 1998).

The mutagenic components in the aqueous extracts of betel quid ingredients were studied by Chen et al. (1984) and Stich et al. (1986), who reported that nitrite-treated aqueous extracts of the fruits and leaves exhibited a mutagenic response only at an acidic pH, and the influence may be neutralized when chewed with lime. Cancers in general are multifactorial in origin, and several environmental interactions are possible; betel chewing may be one of the contributing factors. The reports available on the effects of pan chewing are contradictory.

Betel chewing has been identified as a contributing factor in cancers of the esophagus (Nayar et al. 2000) and mouth (Merchant et al. 2000; Zain et al. 1999) and liver damage (Sarma et al. 1992), but some studies report anti-inflammatory activities (Saeed et al. 1993), chemopreventive activities of betel leaf for skin cancer and breast cancer (Bhide et al. 1994; Murakami et al. 2000; Rao et al. 1985), and activity against tobacco-specific carcinogens (Padma et al. 1989). Topical application of betel leaf extracts inhibited skin cancer in mice by affecting both glutathione and glutathione *S*-transferase levels (Azuine et al. 1991). Leaf and stalk extracts are reported to be antibacterial (Shitut et al. 1999).

REFERENCES

Azuine MA, Amonkar AJ, Bhide SV. 1991. Chemopreventive efficacy of betel leaf extract and its constituents on 7,12-dimethylbenz(a)anthracene induced carcinogenesis and their effect on drug detoxification system in mouse skin. Indian Journal of Experimental Biology 29(4): 346-351.

Balentine DA, Wiseman SA, Bouwens LC. 1997. The chemistry of tea flavanoids. Critical Reviews in Food Science and Nutrition 37: 693-704.

Bhide SV, Azuine MA, Lahiri M, Telang NT. 1994. Chemoprevention of mammary tumor virus-induced and chemical carcinogen-induced rodent mammary tumors by natural plant products. Breast Cancer Research and Treatment 30(3): 233-242.

Chen HC, Chang YS, Lee TC. 1984. The mutagenicity of nitrite-treated aqueous extract of *Piper betle* L. Proceedings of the National Science Council, Republic of China. Part B. Life Sciences 8(1): 4-10.

Dar A, Khatoon S. 2000. Behavioral and biochemical studies of dichloromethane fraction from the *Areca catechu* nut. Pharmacology, Biochemistry and Behavior 65(1): 1-6.

Ghelardini C, Galeotti N, Lelli C, Bartolini A. 2001. M1 receptor activation is a requirement for arecoline analgesia. Il Farmaco 56(5-7): 383-385.

Graham HN. 1992. Green tea composition, consumption, and polyphenol chemistry. Preventive Medicine 21: 334-350.

Hamilton-Miller JM. 1995. Antimicrobial properties of tea (*Camellia sinensis* L.). Antimicrobial Agents and Chemotherapy 39: 2375-2377.

Hamilton-Miller JM. 2001. Anti-carcinogenic properties of tea (*Camellia sinensis*). Journal of Medical Microbiology 50(4): 299-302.

Hegarty VM, May HM, Khaw KT. 2000. Tea drinking and bone mineral density in older women. American Journal of Clinical Nutrition 71: 1003-1007.

Huang CL, Yang CC, Deng JF. 1997. Betel nut poisoning in Taiwan. Toxicon 35(4): 501.

Imai K, Nakachi K. 1995. Cross sectional study of effects of drinking green tea on cardiovascular and liver diseases. British Medical Journal 310: 693-696.

Inokuchi J, Okabe H, Yamauchi T, Nagamatsu A, Nonaka G, et al. 1986. Antihypertensive substance in seeds of *Areca catechu* L. Life Sciences 38(15): 1375-1382.

Jeng JH, Hahn LJ, Lin BR, Hsieh CC, Chan CP, et al. 1999. Effects of areca nut, inflorescence *Piper betle* extracts and arecoline on cytotoxicity, total and unscheduled DNA synthesis in cultured gingival keratinocytes. Journal of Oral Pathology and Medicine 28(2): 64-71.

Kono S, Shinchi K, Ikeda N, Yanai F, Imanishi K. 1992. Green tea consumption and serum lipid profiles: A cross-sectional study in Northern Kyushu. Preventive Medicine 21: 526-531.

Kumar M, Kannan A, Upreti RK. 2000. Effect of betel/areca nut (*Areca catechu*) extracts on intestinal epithelial cell lining. Veterinary and Human Toxicology 42(5): 257-260.

Mannan N, Boucher BJ, Evans SJ. 2000. Increased waist size and weight in relation to consumption of *Areca catechu* (betel-nut): A risk factor for increased glycaemia in Asians in east London. British Journal of Nutrition 83(3): 267-275.

Mehrotra SC, Pande P, Kumar N, Sharma PN. 1999. Mineral composition of leaves of seven betel vine cultivars. Indian Journal of Agricultural Biochemistry 12(1): 36-37.

Merchant A, Husain SS, Hosain M, Fikree FF, Pitiphat W, et al. 2000. Paan without tobacco: An independent risk factor for oral cancer. International Journal of Cancer 86(1): 128-131.

Murakami A, Ali AM, Mat-Salleh K, Koshimizu K, Ohigashi H. 2000. Screening for the *in vitro* anti-tumor-promoting activities of edible plants from Malaysia. Bioscience, Biotechnology, and Biochemistry 64(1): 9-16.

Nayar D, Kapil U, Joshi YK, Sundaram KR, Srivastava SP, et al. 2000. Nutritional risk factors in esophageal cancer. Journal of the Association of Physicians of India 48(8): 781-787.

Noor AM. 1987. Quantitative analysis of areca catechu (betel) nut flavanols (tannins) in relation to oral submucous fibrosis. Dental Journal of Malaysia 9(2): 29-32.

Norton SA. 1998. Betel: Consumption and consequences. Journal of the American Academy of Dermatology 38(1): 81-88.

Padma PR, Lalitha VS, Amonkar AJ, Bhide SV. 1989. Anticarcinogenic effect of betel leaf extract against tobacco carcinogens. Cancer Letters 45(3): 195-202.

Rao AR, Sinha A, Selvan RS. 1985. Inhibitory action of *Piper betle* on the initiation of 7,12-dimethylbenz[a]anthracene-induced mammary carcinogenesis in rats. Cancer Letters 26(2): 207-214.

Saeed SA, Farnaz S, Simjee RU, Malik A. 1993. Triterpenes and B-sitosterol from *Piper betle*: Isolation, antiplatelet and anti-inflammatory effects. Biochemical Society Transactions 21(4): 462S.

Sagesaka-Mitane Y, Miwa M, Okada S. 1990. Platelet aggregation inhibitors in hot water extract of green tea. Chemical and Pharmaceutical Bulletin 38: 790-793.

Sarma AB, Chakrabarti J, Chakrabarti A, Banerjee TS, Roy D, et al. 1992. Evaluation of pan masala for toxic effects on liver and other organs. Food and Chemical Toxicology 30(2): 161-163.

Serafini M, Ghiselli A, Ferro-Luzzi A. 1996. *In vivo* antioxidant effect of green tea in man. European Journal of Clinical Nutrition 50: 28-32.

Sharan RN, Wary KK. 1992. Study of unscheduled DNA synthesis following exposure of human cells to arecoline and extracts of betel nut *in vitro*. Mutation Research 278(4): 271-276.

Shitut S, Pandit V, Mehta BK. 1999. The antimicrobial efficiency of *Piper betle* Linn leaf (stalk) against human pathogenic bacteria and phytopathogenic fungi. Central European Journal of Public Health 7(3): 137-139.

Stensvold I, Tverdal A, Solvoll K, Foss OP. 1992. Tea consumption. Relationship to cholesterol, blood pressure, and coronary and total mortality. Preventive Medicine 21: 546-553.

Stich HF, Rosin MP, Brunnemann KD. 1986. Oral lesions, genotoxicity and nitrosamines in betel quid chewers with no obvious increase in oral cancer risk. Cancer Letters 31(1): 15-25.

Stoner GD, Mukhtar H. 1995. Polyphenols as cancer chemopreventive agents. Journal of Cellular Biochemistry 22: 169-180.

Taylor RF, Al-Jarad N, John LM, Conroy DM, Barnes NC. 1992. Betel-nut chewing and asthma. Lancet 339(8802): 1134-1136.

Wary KK, Sharan RN. 1991. Cytotoxic and cytostatic effects of arecoline and sodium nitrite on human cells *in vitro*. International Journal of Cancer 47(3): 396-400.

Yamaguchi Y, Hayashi M, Yamazoe H, Kunitomo M. 1991. Preventive effects of green tea extract on lipid abnormalities in serum, liver and aorta of mice fed an atherogenic diet. Nippon Yakurigaku Zasshi 97: 329-337.

You SQ. 1993. Study on feasibility of Chinese green tea polyphenols (CTP) for preventing dental caries. Chinese Journal of Stomatology 28: 197-199.

Zain RB, Ikeda N, Gupta PC, Warnakulasuriya S, van Wyk CW, et al. 1999. Oral mucosal lesions associated with betel quid, areca nut and tobacco chewing habits: Consensus from a workshop held in Kuala Lumpur, Malaysia, November 25-27, 1996. Journal of Oral Pathology and Medicine 28(1): 1-4.

Index

Page numbers followed by the letter "t" indicate tables.

doi:10.1300/5504_08

Printed in the United Kingdom
by Lightning Source UK Ltd.
132295UK00002B/154-162/P

9 781560 223122